高等职业教育规划教材

# 通用航空专业英语

Technical English for General Aviation

周谧 主 编
胥郁 副主编

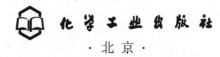

## 内容提要

《通用航空专业英语》共分4个单元，18个章节，主要内容包括通用航空概述、美国通用航空、通用航空飞机及制造和通用航空的应用。每个单元包括Background，General Vocabulary，Notes和Exercises四个部分。本书的设计旨在通过大量的素材阅读、反复使用课文要求词汇，达到掌握一定量通航相关的专业词汇、提高专业阅读和译技能的目的。为方便教学，本书配有电子课件。

本书是通航专业学生在完成基础英语阶段之后的通航专业英语课程教材。本书适合高职院校通用航空航务技术、通用航空器维修、民航运输等专业的学生或具有同等水平的读者参考学习，也适合对通用航空事业感兴趣的读者学习和参考。

**图书在版编目（CIP）数据**

通用航空专业英语/周谧主编．—北京：化学工业出版社，2020.8
高等职业教育规划教材
ISBN 978-7-122-37080-8

Ⅰ.①通… Ⅱ.①周… Ⅲ.①航空-英语-高等职业教育-教材 Ⅳ.①V2

中国版本图书馆CIP数据核字（2020）第085959号

责任编辑：旷英姿　王　可　　　装帧设计：王晓宇
责任校对：赵懿桐

出版发行：化学工业出版社（北京市东城区青年湖南街13号　邮政编码100011）
印　　刷：三河市航远印刷有限公司
装　　订：三河市宇新装订厂
787mm×1092mm　1/16　印张16½　字数421千字　2020年10月北京第1版第1次印刷

购书咨询：010-64518888　　　售后服务：010-64518899
网　　址：http://www.cip.com.cn
凡购买本书，如有缺损质量问题，本社销售中心负责调换。

定　价：45.00元　　　　　　　　　　　　　　　　　版权所有　违者必究

# 前言
PREFACE

近年来,随着我国经济的快速发展,人们对发展通用航空的需求日益高涨。特别随着国务院和中央军委联合下发《关于深化我国低空空域管理改革的意见》及《国家中长期科技发展规划纲要(2006—2020)》,我国低空空域管理改革的大幕拉开,通用航空产业发展迎来了历史性的发展机遇。

为了有力推动我国通用航空事业的健康发展,我们需要广泛汲取美国等通航发达国家的成功经验,进而系统把握通用航空体系发展全貌。然而,目前关于介绍通用航空的专业英文教材甚少,因此,从专业英语角度,编写一本系统介绍通用航空的英文教材十分必要。本书主要内容包括通用航空现状、通用航空与低空空域、通用航空机场、中国通用航空、美国航空小镇、飞行文化、美国通用航空机场建设、美国通用航空现状、通用航空飞机机型分类、主要通用航空飞机制造商、公务机、小型飞机、通用航空航空器选购、租赁和维修、医疗救援、农业航空、空中游览、工业航空和飞行执照等内容。本书以学生为中心,采取启发式任务型教学模式,通过课上引导以及进行大量的阅读练习来拓宽学习者的专业知识面,提高学习者的阅读质量,丰富阅读技巧,培养推理概括能力。课后练习围绕每章主题,采用分段式介绍,以期增加学习者的阅读范围,扩大词汇量,加强句子结构的运用并能表现在口语表达上,从而提高学习者的英语水平。

本书由周谧主编,胥郁副主编,全书由周谧负责统稿。具体编写分工如下:第一单元由长沙航空职业技术学院胥郁编写;第二单元由长沙航空职业技术学院何裕龙和易夫编写;第三、第四单元由长沙航空职业技术学院周谧编写。本书的编写还得到了民航湖南监管局通用航空处的大力支持,为本书编写提供了难得的素材和宝贵的建议。

本书在编写过程中,广泛采纳和吸取了国内很多专家学者和业内资深人士的观点,参考了相关书籍和文章,在此表示衷心感谢。由于我国通用航空发展很快,市场环境和政策环境变化很大,许多问题还在不断实践和探讨中,难以形成定论;加之编者水平有限,书中难免存在疏漏和不足之处,敬请各位读者不吝指教。

编者
2020年4月

# 目 录 CONTENTS

通用航空专业英语

## Unit 1 General introduction of GA 通用航空概述 /001

1.1 Status quo of general aviation 通用航空现状 /002
1.2 GA and low-altitude airspace 通用航空与低空空域 /012
1.3 GA airport 通用航空机场 /021
1.4 GA in China 中国通用航空 /030

## Unit 2 General Aviation in USA 美国通用航空 /042

2.1 Aviation town in the United States 美国航空小镇 /043
2.2 Aviation culture 飞行文化 /054
2.3 GA airport in USA 美国通用航空机场建设 /062
2.4 Status of GA in USA 美国通用航空现状 /073

## Unit 3 GA aircrafts and manufacture 通用航空飞机及制造 /083

3.1 Classifications of GA aircraft 通用航空飞机机型分类 /084
3.2 Major manufacturers of GA aircraft 主要通用航空飞机制造商 /096
3.3 Business aircraft 公务机 /107
3.4 Small aircraft 小型飞机 /118
3.5 The procurement, leasing and maintenance of GA aircrafts 通用航空航空器选购、租赁和维修 /131

## Unit 4 Applications of GA 通用航空的应用 /142

4.1 Emergency medical service 医疗救援 /143
4.2 Agricultural aviation 农业航空 /152
4.3 Air tourism 空中游览 /165

4.4　Industrial aviation 工业航空　/173
4.5　Pilot license 飞行执照　/184

# Key to Exercises 练习答案　/196

# Vocabulary 词汇表　/215

# 相关通用航空知识　/252

# 参考文献　/257

# Unit 1

# General introduction of GA
# 通用航空概述

# 1.1 Status quo of general aviation
## 通用航空现状

## Background

General Aviation (hereinafter GA) refers to the civil aviation activities by the use of civil aircraft engaging not in public air transport but in air operations for works of industry, agriculture, forestry, fishery and construction and other flight activities for medical and health care, emergency rescue and disaster relief, meteorological observation, maritime monitoring, scientific experiments, and programs of education, training, culture and sports. Thanks to the rapid development of aviation technology and the conversion of a large number of military aircraft to civilian use after Word War II, the GA attained a rapid growth and ushered in its booming ages in the 1970s. In the 1980s, however, due to the global economic slowdown, GA was frustrated by the relatively saturate number of aircraft together with the fall in technological innovation, and consequently GA began to decline and slipped in the doldrums. With the continuous growth of world economy in the 1990s, and as the governments around the world introduced encouraging policies and new aviation products were launched, GA saw its revitalization and rise again with great and strong potentials.

（1）GA in China

In recent years, general aviation has been developing at fast pace in China, with its annual total flights growing on average by above 10%. And specifically it has been seeing its business scaling up on a gradual basis, fields of application expanding, and its flights increasing in both varieties and demands. Now as the economy continues to grow rapidly and the living standards of urban and rural residents keep improving, it is expected that in the following decade average annual GA growth rate in China will sit at more than 15%. Nevertheless, Chinese GA industry is just getting started, making it difficult to satisfy the needs of national economy growth, and to provide due support for China's economic development and transformation. And how to promote GA's development turns out to be a major topic worthy of studies. The three major issues stumbling the development of GA in China are inefficient access to airspace, and severe shortage of GA airports and pilots.

Overall, the general aviation in China has maintained a sustained and rapid development. There're six indices reflecting the basic status quo of the development of GA in a country, namely the number of GA enterprises, annual flight (air operation) hours, GA's fleet size, number of GA airport, number of people engaging in GA, and social and economic benefits. With respect to air operation hours, the general aviation operations completed for the production of all lines of industries totaled 517,000 hours in 2012, a year-on-year increase of 2.8%. Among them, industrial aviation operation accounted for 77,100 hours, up 36% from a year earlier, agricultural and forestry aviation operation for 31,900 hours, and other GA operations for a sum of 408,100 hours.

With respect to the capability and level of disaster relief and accident rescue, general aviation has played an irreplaceable role to which other modes of transportation can't contribute. In 2008, GA had fully proven its importance in the disaster relief actions in the earthquake that devastated Wenchuan County, under the condition of land transportation being cut off, the way to transfer the wounded and transport goods at that very moment rested only on the helicopters. As for the capability and level of operations in industrial aviation and agricultural and forest aviation, it was at fairly early stage of the developmental history of GA in China that air operations were applied in agriculture, forestry and industry, whereas the total of their combined operation hours were less than 110,000 hours in 2012, which showed that GA's development in these areas was slow and not widely expanded. When it comes to the business flight and private flight of residents, currently there are 19 established GA enterprises plus nearly 50 more enterprises planned to be established, with their main scope of business on provision of business flight and administration of business aircrafts on other's behalf. At the end of the year of 2012, the fleet of business aircrafts consisted of 166 aircrafts, and it was estimated that such fleet would be expanded to more than 200 in size in 2013, the majority of which would be business jets in various types.

(2) GA in United States

Today, general aviation in United States has been so developed that it, together with the transport aviation, comprises the country's safest and most efficient air transportation system. It holds together the overall balance of civil air transportation industry, and bridges United States with the rest of the world. GA also lays the foundation for air transport. In America, aviation services are convenient and practical for almost everyone, because transport aviation carriers and GA companies play a fully separately independent but collectively harmonious role in transportation service. GA's presence is mainly outside the metropolis, and with its flexible and specialized methods, it transports annually millions of passengers as well as tens of thousands of tons of cargo and it delivers mails and packages to more remote regions at the shortest time. Meanwhile, flights of GA are also operated every now and then in the airports of certain major cities in the United States.

In the United States, a country where general aviation is most developed, GA provides both American government and general public with services ranging from emergency medical flights, important aerial law enforcement support (such as customs and border protection, forest services, etc), disaster relief, search and rescue to flight training, and also it supplies in bulk other services such as chartered passenger service, official flight, self-piloting business flight and air freight service, which have greatly facilitated the travel of residents in remote areas. To be noted, many of the aviation services provided by GA are not effectively provided by commercial service airports.

### (3) GA in United Kingdom

General aviation plays as an essential role as roads, railways, airlines and waterways in a country's transportation system. And GA in the United Kingdom mainly provides support for emergency service. With more than 1000 helicopters registered for civil use in the country, the majority of them act as an essential role in the frontier areas of emergency services (ambulance, security, firefighting, search and rescue) or support economic tasks as necessary. And helicopters are also indispensable in the offshore oil and natural gas industry, which together account for 2.5% of the UK's GDP. Other roles engaged by GA include aviation surveys, patrol inspections, film-making, lifting and passenger transportation. Modern commercial helicopters are safe, reliable, and with significantly lower noise.

The British liken the general aviation aircraft to a winged car that like the car, can be used for business and entertainment. It is understood that GA's development is closely dependent on the layout of airport and the smoothness of airspace traffic, and if any of which is improperly restricted, it will substantially undermine the development of general aviation and regional economy growth. Hence, just as the publically accepted saying explains, *a mile of highway will take you just one mile, but a mile of runway will take you anywhere.*

All in all, the general aviation industry in China has long been making significant contributions to supporting industrial and agricultural production, protecting people's lives and property, and preventing against various natural disasters. It has achieved marked social benefits, but failed to attain satisfactory economic returns. The status quo of GA's development in China can be summarized in three points: first, it starts up late, second, it grows slowly, and third, it is small in size and scale. Comparatively speaking, China, as a large country in agriculture and tourism and a country powered with rapid economic growth, will see its general aviation market outperformed along with its gradual easing of restrictions in policies, regulations and capitals. And indeed there is great room and potential to tap into China's GA industry and it is just a matter of time before it develops into the form of the one in the United States.

## General Vocabulary

General Aviation (GA)　通用航空

# Unit 1  General introduction of GA 通用航空概述

industry    *n.* 产业；工业；勤勉
agriculture    *n.* 农业，农耕；农业生产；农艺，农学
forestry    *n.* 林业；森林地；林学
fishery    *n.* 渔业；渔场；水产业
construction    *n.* 建设；建筑物；解释；造句
health care    医疗卫生
emergency rescue    急救
disaster relief    救灾
meteorological observation    气象观测
maritime monitor    海洋监测
military aircraft    军用飞机
civilian    *adj.* 民用的，百姓的，平民的
attain    *vt.* 达到，实现；获得；到达
usher    *vt.* 引导，招待；迎接；开辟
pace    *n.* 一步；步速；步伐；速度
scale up    按比例放大；按比例增加
expand    *vt.* 扩张；使膨胀；详述
gradual    *adj.* 逐渐的；平缓的
living standard    生活水平，生活标准
decade    *n.* 十年，十年期；十
nevertheless    *adv.* 然而，不过；虽然如此
transformation    *n.* 转化；转换；改革；变形
promote    *vt.* 促进；提升；推销；发扬
slowdown    *n.* 减速；怠工；降低速度
frustrate    *vt.* 挫败；阻挠；使感到灰心
saturate    *adj.* 浸透的，饱和的；深颜色的
technological innovation    技术革新；工艺革新
consequently    *adv.* 因此；结果；所以
decline    *vi.* 下降；衰落；谢绝
doldrums    *n.* 低谷，忧郁；赤道无风带
encourage    *vt.* 鼓励，怂恿；激励；支持
launch    *v.* 发射（导弹、火箭等）；发起，发动；使……下水；开始；起飞
revitalization    *n.* 复兴，复苏
potential    *n.* 潜能，可能性；电势
average    *adj.* 平均（数）的；普通的；典型的；平庸的；中等的，适中的
application    *n.* 应用；申请；应用程序；敷用；（对事物、学习等）投入
demand    *n.* （坚决的或困难的）要求；（顾客的）需求
urban    *adj.* 城市的；住在都市的
kingdom    *n.* 王国；（某人）占统治地位的地方；管辖范围；（某种特质为重的）领域
rural    *adj.* 农村的，乡下的；田园的，有乡村风味的

resident　n.居民；（旅店）房客；住院医生，（美国的）高级专科住院实习医生；〔美〕寄宿制学校学生；（英国派驻半独立国家的）特派代表；驻外特工

worthy　adj.值得的；有价值的；配得上的，相称的；可尊敬的；应……的

stumble　vt.使……困惑；使……绊倒

inefficient　adj.无效率的，效率低的；无能的

airspace　n.空域；领空；空间

shortage　n.缺乏，缺少；不足

maintain　vt.维持；继续；维修；主张；供养

sustain　vt.维持；支撑，承担；忍受；供养；证实

indice　n.指数；标记体

status quo　现状

namely　adv.也就是；即是；换句话说

enterprise　n.企业；事业；进取心；事业心

fleet　n.船队，舰队；捕鱼船队；（一国的）海军；（同一机构的）车队，机群；（沼泽地）小河，水道，沟

engage in　从事于（参加）

benefit　n.利益，好处；救济金

account for　对……负有责任；对……做出解释；说明……的原因；导致；（比例）占

year-on-year　adj.与上年同期数字相比的

sum　n.金额；总数

capability　n.才能，能力；性能，容量

rescue　n.营救，解救，援救；营救行动

irreplaceable　adj.（因贵重或独特）不能替代的，独一无二的；失掉（或损伤）后无法补偿的

earthquake　n.地震；大动荡

devastate　vt.毁灭；毁坏

land transportation　陆路运输

helicopter　n.直升机

fairly　adv.相当地；公平地；简直

establish　v.建立，创立；确立；获得接受；查实，证实

provision　n.规定；条款；准备；供应品

consist of　由……组成；由……构成；包括

estimate　v.估计，估量；判断，评价

efficient　adj.有效率的；有能力的；生效的

foundation　n.基础；地基；基金会；根据；创立

convenient　adj.方便的；适当的；近便的；实用的

harmonious　adj.和谐的，和睦的；协调的；悦耳的

flexible　adj.灵活的；柔韧的；易弯曲的

law enforcement　法律的实施；执法机关

charter　　v. 包租；发给特许执照
air freight　　航空运费；空运的货物
railway　　n.〔英〕铁路；轨道；铁道部门
frontier　　adj. 边界的；开拓的
GDP　　abbr. 国内生产总值（gross domestic product）
reliable　　adj. 可靠的；可信赖的
noise　　n. 噪声；响声；杂音
liken to　　与……相比，把……比作
entertainment　　n. 娱乐；消遣；款待
dependent　　adj. 依靠的；从属的；取决于……的
layout　　n. 布局；设计；安排；陈列
restrict　　vt. 限制；约束；限定
substantially　　adv. 实质上；大体上；充分地
undermine　　vt. 破坏，渐渐破坏；挖掘地基
mile　　n. 英里；一英里赛跑；较大的距离
contribution　　n. 贡献；捐献；投稿
protect　　vt. 保护，防卫；警戒
property　　n. 性质，性能；财产；所有权
summarize　　vt. 总结；概述
ease　　v. 减轻，缓解；小心缓缓地移动；使容易；放松；（使）贬值；（股票价格、利率等）下降，下跌

## Notes

1. General Aviation (hereinafter GA) refers to the civil aviation activities by the use of civil aircraft engaging not in public air transport but in air operations for works of industry, agriculture, forestry, fishery and construction and other flight activities for medical and health care, emergency rescue and disaster relief, meteorological observation, maritime monitoring, scientific experiments, and programs of education, training, culture and sports.

– refer to　　参考；涉及；指的是；适用于
– air transportation　　航空运输

译文：通用航空，是指使用民用航空器从事公共航空运输以外的民用航空活动，包括从事工业、农业、林业、渔业和建筑业的作业飞行以及医疗卫生、抢险救灾、气象探测、海洋监测、科学实验、教育训练、文化体育等方面的飞行活动。

2. With the continuous growth of world economy in the 1990s, and as the governments around the world introduced encouraging policies and new aviation products were launched, GA saw its revitalization and rise again with great and strong potentials.

– encouraging policy　　鼓励政策

— continuous growth 持续增长

译文：20世纪90年代以来，随着世界经济的持续增长、各国政府出台鼓励政策、航空产品推陈出新，通用航空又呈现复苏和重新崛起的态势，且有强劲后势。

3. In recent years, general aviation has been developing at fast pace in China, with its annual total flights growing on average by above 10%. And specifically it has been seeing its business scaling up on a gradual basis, fields of application expanding, and its flights increasing in both varieties and demands.

— fast pace 快节奏

— on a gradual basis 逐渐地

— varieties and demands 种类和需求

译文：近年来，我国通用航空快速发展，飞行总量年均增长达10%以上，行业规模日益扩大，应用领域不断拓展，飞行种类日益增多，飞行需求渐趋旺盛。

4. Now as the economy continues to grow rapidly and the living standards of urban and rural residents keep improving, it is expected that in the following decade average annual GA growth rate in China will sit at more than 15%.

— living standard 生活水平，生活标准

— urban and rural residents 城乡居民

— It is expected that… 预计，估计

— growth rate 增长速率

译文：随着经济持续快速发展和城乡居民生活水平的不断提高，预计今后10年间我国通用航空年均增长将达到15%以上。

5. The three major issues stumbling the development of GA in China are inefficient access to airspace, and severe shortage of GA airports and pilots.

— major issue 主要问题

— access to 接近；有权使用；通向……的入口

— airspace 空域

— severe shortage 严重短缺

译文：空域无法有效使用、通用机场严重不足以及飞行员严重短缺成为阻碍我国通用航空发展的最主要的三大障碍。

6. With respect to the capability and level of disaster relief and accident rescue, general aviation has played an irreplaceable role to which other modes of transportation can't contribute.

— with respect to 关于，至于

— disaster relief 抢险救灾

— accident rescue 事故救援

— play a role 发挥作用；扮演一个角色

译文：从抢险救灾与事故救援的能力和水平来看，通用航空发挥了其他运输方式不可替代的作用。

7. As for the capability and level of operations in industrial aviation and agricultural and forest aviation, it was at fairly early stage of the developmental

history of GA in China that air operations were applied in agriculture, forestry and industry, whereas the total of their combined operation hours were less than 110,000 hours in 2012, which shows that GA's development in these areas is slow and not widely expanded.

- industrial aviation　工业航空，包括使用航空器进行工矿业有关的各种活动，具体的应用有航探、航空吊装、石油航空、航空环境监测等。
- early stage　早期，初期
- whereas　然而；鉴于

译文：从工业航空作业和农林作业能力和水平看，我国通用航空发展过程中，农林作业和工业作业发展较早，但2012年二者年作业小时之和不到11万小时，可见其发展之缓慢，也说明应用领域不广。

8. GA also lays the foundation for air transport. In America, aviation services are convenient and practical for almost everyone, because transport aviation carriers and GA companies play a full separately independent but collectively harmonious role in transportation service.

- lay the foundation　奠定基础
- transport aviation carriers　运输航空公司，carrier　承运人
- transportation service　运输服务

译文：通用航空是航空运输的基础，在美国，航空服务几乎对每一个人都便利、实用，因为运输航空公司和通用航空公司充分发挥各自独立但和谐的运输作用。

9. GA's presence is mainly outside the metropolis, and with its flexible and specialized methods, it transports annually millions of passengers as well as tens of thousands of tons of cargo and it delivers mails and packages to more remote regions at the shortest time.

- metropolis　大都市；首府；重要中心
- as well as　等同于and
- cargo　货物
- remote regions　偏远地区

译文：通用航空主要活跃在大都市以外，以其机动灵活的特有方式每年运输百万旅客、千万吨货物，以最快的速度将邮件送达更偏远的地区。

10. In the United States, a country where general aviation is most developed, GA provides both American government and general public with services ranging from emergency medical flights, important aerial law enforcement support (such as customs and border protection, forest services, etc), disaster relief, search and rescue to flight training, and also it supplies in bulk other services such as chartered passenger service, official flight, self-piloting business flight and air freight service, which have greatly facilitated the travel of residents in remote areas.

- range from...to...　从……到……变动，从……到……范围
- law enforcement　执法机关
- in bulk　整批，散装；大批，大量

- charter service　包机服务
- air freight　航空货运

译文：在通用航空最为发达的美国，通用航空为美国政府和公众提供了紧急医疗飞行、重要的航空执法服务（如海关和边境保护、森林服务等）、抢险救灾和搜救、飞行训练等服务，也提供了大量包机客运服务、公务飞行服务、自驾商务飞行、航空货运服务，为偏远地区的居民出行提供了极大便利。

11. And helicopters are also indispensable in the offshore oil and natural gas industry, which together accounts for 2.5% of the UK's GDP. Other roles engaged by GA include aviation surveys, patrol inspections, film-making, lifting and passenger transportation. Modern commercial helicopters are safe, reliable, and with significantly lower noise.

- offshore oil　近海石油
- account for　（比例）占

译文：直升机在海洋石油和天然气工业中是必不可少的，而这两大工业占了英国GDP(国内生产总值)的2.5%，其他各种各样的角色包括了航空测量、巡查、拍电影、起重和旅客运输。现代商业直升机安全可靠，噪声也明显越来越小。

12. All in all, the general aviation industry in China has long been making significant contributions to supporting industrial and agricultural production, protecting people's lives and property, and preventing against various natural disasters. It has achieved marked social benefits, but failed to attain satisfactory economic returns.

- make contribution to　为……作贡献
- prevent against　防止，防范
- economic return　经济效益，经济回报

译文：总之，我国通用航空长期以来为支持工农业生产、保护人民生命财产和预防、抵御各种自然灾害做出了重要贡献，社会效益比较显著，但经济效益并不明显。

## Exercises

Section one: Answer the following questions according to the text.

1. What is the general aviation?
2. What services does GA include?
3. What are the differences between GA in United States and GA in United Kingdom?
4. Compared to GA in the US, what problems does GA in China encounter?
5. What are the advantages or potentials of GA market in China?

Section two: Match the words or phrases in column A with the definition in column B.

| A | B |
|---|---|
| 1. civilian | a. something that serves as a means of transportation |
| 2. military | b. something that aids or promotes well-being |
| 3. freight | c. extend in one or more directions |
| 4. transport | d. nonmilitary citizen |
| 5. expand | e. goods carried by a large vehicle |
| 6. benefit | f. the military forces of a nation |
| 7. satisfactory | g. keep from happening or arising; make impossible |
| 8. prevent | h. meeting requirements |
| 9. potential | i. being effective without wasting time or effort or expense |
| 10. efficient | j. the inherent capacity for coming into being |

Section three: Translate the following into Chinese.

1. Today, general aviation in United States has been so developed that it, together with the transport aviation, comprises the country's safest and most efficient air transportation system.

2. Now as the economy continues to grow rapidly and the living standards of urban and rural residents keep improving, it is expected that in the following decade average annual GA growth rate in China will sit at more than 15%.

3. In 2008, GA has fully proven its importance in the disaster relief actions in the earthquake that devastated Wenchuan County, under the condition of land transportation being cut off, the way to transfer the wounded and transport goods at that very moment rested only on the helicopters.

4. Flights of GA are also operated every now and then in the airports of certain major cities in the United States.

5. With more than 1000 helicopters registered for civil use in the country, the majority of them acts as the necessary role in the frontier areas of emergency services (ambulance, security, firefighting, search and rescue) or supports economic tasks as necessary.

6. It holds together the overall balance of civil air transportation industry, and bridges United States with the rest of the world.

7. It is understood that GA's development is closely dependent on the layout of airport and the smoothness of airspace traffic, and if any of which is improperly restricted, it will substantially undermine the development of general aviation and regional economy growth.

8. And indeed there is a great room and potential to tap in China's GA industry and it is just a matter of time before it develops into the form of the one in the United States.

Section four: Translate the following into English.
1. 所有捐款，无论多少，我们都深表感谢。
2. 在你的手提箱上贴上标签以防弄混。
3. 自从这个国家的两大支柱产业衰退后，其经济一直萎靡不振。
4. 政府的经济政策使我们遭遇了多年来最为严重的经济衰退。
5. 炸弹炸死4名士兵及3名平民。
6. 在春天和初夏之时植物长势最为明显。
7. 电子出版业是一个正在发展壮大的产业。
8. 学会如何应付突发事件是海外旅游的众多益处之一。
9. 通过事后反省，我们比较容易认清自己所犯的错误。
10. 我们将聘用专业管理人员为我们服务。

# 1.2
# GA and low-altitude airspace
## 通用航空与低空空域

## Background

As an integral part of a country's transport and economic infrastructure, general aviation can enhance economic and social development, integrate remote areas, connect markets and communities, and make millions of employment opportunities available directly and indirectly, generating billions worth of economic benefits. However, due to the limitations by complex airspace structure and inconvenient approval of flight schedules, China's potential demand for GA has long been untapped. And the unavailability to operate flight and access to smooth airway turns out to be the key bottleneck in the development of general aviation in China.

## Status quo of low-altitude airspace

Low-altitude airspace is the main space for general aviation activities, and safe and convenient airspace resources as such are the preset condition for the thriving development of general aviation industry. Earlier, China's low-altitude airspace has been kept under control and its management reform lagged behind, which pose difficulties for GA planes in both taking off and landing. Hence gradual and increasing opening-up of low-altitude airspace will be an important measure to improve GA's development. Reform on low-altitude airspace management in China will be extended from 1000 meters to 3000 meters so as to realize gap-free connection between the surveillance airspace and the report airspace.

In China, airspaces for civil aviation purpose are confined to airways, air routes and other airspace around civil airports, at merely a sum of 10% of the total airspaces. Apart from that, all remaining airspaces are fallen under and administrated by military aviation to conduct training, patrolling and etc, and the low-altitude airspace needed for general aviation operation is also regulated and controlled by military aviation and to a highly strict extent, making such low-altitude airspace inaccessible at all. As far as the number of airports, enterprises, aircrafts and flight hours for general aviation are concerned, GA in China has made far less progress as opposed to that in developed countries such as USA and Europe. The wide-scale commercialization of GA in China has also not been formally set out, the only large industry sector yet to be opened up since the country's reform and opening-up strategy.

### Low-altitude tourism

Low-altitude tourism refers to people's travel, entertainment and sports in low-altitude airspace through GA transportation, general aircraft and low-flying aircraft. China strongly encourages the development of tourism and sets to exert active efforts to promote low-altitude flight tourism.

To protect the ecological environment, developed countries such as Europe, America and Australia choose not to establish ground transportation facilities in ecologically fragile areas but to develop low-altitude tours instead, such as helicopter tour, which does not necessitate the prior establishment of runway, large parking apron, highways and railways, nor does cause land requisition and destruction to vegetative cover on earth surface, and on the other hand, is environmentally friendly with little impact on surrounding environment, and is convenient, fast, time-saving and less susceptible to the landform while tourists relocate to sightsee, satisfying tourist's in-the-scene, customized and high-end demands.

Low-altitude tourism has attracted the attention of the governments of developed countries. Many countries have formulated long-term development strategies and corresponding policies, invested funds, trained relevant professionals, and established a sound law, regulation and standard system as well as GA industry organizations, which altogether put the whole international low-altitude tourism on the track of fostering a variety of industrial businesses with the core theme of low-altitude tourism. For instance, Niagara Falls Park offers helicopter and hot-air balloon tours; every day in Colorado Canyon, dozens of helicopters take off one after another to render air cruise service to tourists; Helicopter sightseeing in Rotorua City of New Zealand tours across the Taravila Mountains to demonstrate the wonderful natural volcanic landscape of Rotorua; Brazil's Iguazu Falls, listed as a World Heritage Site by UNESCO can be appreciated with a panoramic view through helicopters; Tourists for Australian Great Barrier Reef, by the ride of helicopter, pass across the Coral Sea to dive, boat, and enjoy the wonderful experience.

Low-altitude tourism, with its low demand for natural resources, caters to the requirement of harmonious development between tourism and ecological conservation, and plays a significant role in developing a sustainable tourism industry. In China, damages to the ecological environment take place again and again in the cause of thriving tourism, and in particular some areas suffer from irreversible ecological disasters. Therefore it's bound to see a new ecologically and environmentally friendly tourism industry burgeoning.

As always, the market has had great expectations for the opening-up of China's low-altitude airspace, and from the perspective of future reform tendency, it is already a common view among management teams to give place to the development of general aviation. In the future, if the low-altitude airspace management reform is fully introduced, GA will be put on a faster track of development, driving growth of sectors such as GA manufacturing, air traffic control, operation, maintenance, training, etc., and hopefully benefiting relevant companies in the industry chain.

## General Vocabulary

integral　*adj.* 积分的；完整的，整体的；构成整体所必须的
infrastructure　*n.* 基础设施；公共建设；下部构造
enhance　*vt.* 提高；加强；增加
community　*n.* 社区；群落；共同体；团体
employment　*n.* 使用；职业；雇用
opportunity　*n.* 时机，机会
due to　由于；应归于
complex　*adj.* 复杂的；合成的
approval　*n.* 批准；认可；赞成
untapped　*adj.* 未开发的；未使用的；塞子未开的
bottleneck　*n.* 瓶颈；障碍物
altitude　*n.* 高地；高度；顶垂线；（等级和地位等的）高级；海拔
preset　*adj.* 预先装置的，预先调整的
thriving　*adj.* 欣欣向荣的，兴旺发达的
lag　*vi.* 滞后；缓缓而行；蹒跚
take off　起飞；脱下；离开
opening-up　*n.* 开放
measure　*n.* 测量；措施；程度；尺寸
improve　*vt.* 改善，增进；提高……的价值
surveillance　*n.* 监督；监视
civil aviation　民用航空
confine　*vt.* 限制；禁闭；（因疾病、残疾）无法离开（床、家、轮椅）；将（自己或自己的活动）局限于；

merely　*adv.* 仅仅，只不过；只是
apart from　远离，除……之外；且不说；缺少
administrate　*vt.* 管理；经营，实施
patrol　*vt.* 巡逻；巡查
regulate　*vt.* 调节，规定；控制；校准；有系统地管理
as far as　至于，直到，远到；就……而言
progress　*n.* 进步，发展；前进
developed country　发达国家
commercialization　*n.* 商品化，商业化
reform and opening-up strategy　改革开放战略
transportation　*n.* 运输；运输系统；运输工具；流放
ecological environment　生态环境
fragile　*adj.* 脆的；易碎的
necessitate　*vt.* 使成为必需，需要；迫使
runway　*n.* 跑道；河床；滑道
vegetative　*adj.* 植物的；植物人状态的，无所作为的；促使植物生长的；有生长力的
destruction　*n.* 破坏，毁灭；摧毁
environmentally friendly　保护生态环境的；对生态环境无害的
sightsee　*vt.* 观光；游览
customize　*vt.* 定做，按客户具体要求制造
high-end　*adj.* 高端的；高档的
attract　*vt.* 吸引；引起
formulate　*vt.* 规划；用公式表示；明确地表达
sound　*adj.* 合理的；无损的；有能力的；充足的；彻底的；熟睡的；资金充实的；（非正式）非常棒的；严厉的；健全的
foster　*v.* 促进；抚育（他人子女一段时间）；收养；把（孩子）交托给养父母
Niagara Falls Park　尼亚加拉瀑布公园
hot-air balloon　热气球
Colorado Canyon　科罗拉多大峡谷
render　*v.* 致使；提供，回报；援助；提交，提出；作出（裁决）；放弃；表达；演奏；翻译；绘制；粉刷；熔化；从（动物身体）提取（蛋白质）；秘密偷渡
cruise　*n.* 乘船游览，游船度假；巡航，巡游
Rotorua City　罗托鲁瓦市
demonstrate　*vt.* 证明；展示；论证
landscape　*n.* 风景；风景画；景色；山水画；乡村风景画；地形
Iguazu Falls　伊瓜苏大瀑布（巴西旅游景点）
world heritage site　世界遗产地
UNESCO　*abbr.* 联合国教科文组织（United Nations Educational, Scientific, and Cultural Organization）
panoramic　*adj.* 全景的

Great Barrier Reef 大堡礁（澳大利亚）
coral sea 珊瑚海
cater to 迎合；为……服务
sustainable *adj.* 可以忍受的；足可支撑的；养得起的；可持续的
suffer from 忍受，遭受；患……病；受……之苦
burgeon *vi.* 萌芽，发芽；迅速增长
tendency *n.* 倾向，趋势；癖好
air traffic control 空中交通管制
maintenance *n.* 维护，维修；保持；生活费用

## Notes

1. As an integral part of a country's transport and economic infrastructure, general aviation can enhance economic and social development, integrate remote areas, connect markets and communities, and make millions of employment opportunities available directly and indirectly, generating billions worth of economic benefits.

— economic infrastructure 经济基础设施

— remote areas 偏远地区

— employment opportunities 就业机会

译文：通用航空是一个国家的运输和经济基础设施的组成部分。可以提升经济与社会发展、整合偏远地域、连接市场和社区，以及提供数百万的直接和间接就业机会，产生数十亿元的经济效益。

2. And the unavailability to operate flight and access to smooth airway turns out to be the key bottleneck in the development of general aviation in China.

— turn out to 结果是

— key 主要的，重要的

— bottleneck 瓶颈

译文："飞不起来""飞不顺畅"是我国通用航空发展的关键瓶颈。

3. Low-altitude airspace is the main space for general aviation activities, and safe and convenient airspace resources as such are the preset condition for the thriving development of general aviation industry.

— low-altitude airspace 低空空域

— thriving development 蓬勃发展

译文：低空空域是通用航空活动的主要区域，安全便捷的低空空域资源，是通用航空产业繁荣发展的前提。

4. Earlier, China's low-altitude airspace has been kept under control and its management reform lagged behind, which pose difficulties for GA planes in both taking off and landing.

— under control 处于控制之下；情况正常

— lag behind 落后；拖欠

译文：之前我国低空空域一直处于管制状态，我国低空空域管理改革缓慢，导致通用航空"起飞难，落地难"。

5. Apart from that, all remaining airspaces are fallen under and administrated by military aviation to conduct training, patrolling and etc, and the low-altitude airspace needed for general aviation operation is also regulated and controlled by military aviation and to a highly strict extent, making such low-altitude airspace inaccessible at all.

- apart from　　除……之外；且不说；并且
- to a highly strict extent　　在很大程度上

译文：其他的航域均被划分给军航用作训练、巡逻等，由军航统一进行管理。其中，通航所需要的低空空域管制权也属于军航，管制非常严格，使得低空资源根本无法使用。

6. To protect the ecological environment, developed countries such as Europe, America and Australia choose not to establish ground transportation facilities in ecologically fragile areas but to develop low-altitude tours instead, such as helicopter tour, which does not necessitate the prior establishment of runway, large parking apron, highways and railways, nor does cause land requisition and destruction to vegetative cover on earth surface, and on the other hand, is environmentally friendly with trivial impact on surrounding environment.

- ecological environment　　生态环境
- ground transportation　　地面运输
- environmentally friendly　　环保

译文：欧美及澳大利亚等发达国家为保护生态环境，放弃在生态脆弱地区建地面交通设施，转而发展低空旅游，如直升机旅游。直升机旅游不需建跑道和大型停机坪，也无须先行建成公路和铁路，不需征用土地，不破坏地表植被，绿色环保，对周围环境影响小。

7. Many countries have formulated long-term development strategies and corresponding policies, invested funds, trained relevant professionals, and established a sound law, regulation and standard system as well as GA industry organizations, which altogether put the whole international low-altitude tourism on the track of fostering a variety of industrial businesses with the core theme of low-altitude tourism.

- long-term　　长期的，从长远来看
- a variety of　　种种；各种各样的

译文：各国纷纷制订了长远发展战略、相应政策并投入资金；建立了完善的法规和标准体系并培育了相关专业人才，建立通用航空行业组织机构，使得整个国际低空旅游的发展形成了众多以低空旅游为核心主题的产业门类。

8. Low-altitude tourism, with its low demand for natural resources, caters to the requirement of harmonious development between tourism and ecological conservation, and plays a significant role in developing a sustainable tourism industry.

- natural resources　　自然资源，天然资源
- ecological conservation　　生态保护

译文：低空旅游以其自然资源需用度小的优势，迎合了旅游与生态和谐发展的需求，对建设可持续发展的旅游业起到很大作用。

9. As always, the market has had great expectations for the opening-up of China's low-altitude airspace, and from the perspective of future reform tendency, it is already a common view among management teams to give place to the development of general aviation.

— opening-up　开放，扩大开放；疏通，打通

— from the perspective of　从……的角度来看

— common view　普遍的观点；共识

译文：市场对中国低空空域开放一直抱有较大预期，且从未来改革的趋势来看，发展通用航空已成为管理层的共识。

10. In the future, if the low-altitude airspace management reform is fully introduced, GA will be put on a faster track of development, driving growth of sectors such as GA manufacturing, air traffic control, operation, maintenance, training, etc., and hopefully benefiting relevant companies in the industry chain.

— air traffic control（ATC）　空中交通管制

— industry chain　产业链

译文：若未来全面推开低空空域管理改革，将加速推进通用航空产业发展，并带动通航制造、空管、运营、维修、培训等各环节发展，相关产业链公司也有望直接受益。

11. The wide-scale commercialization of GA in China has also not been formally set out, the only large industry sector yet to be opened up since the country's reform and opening-up strategy.

— wide-scale　大规模的，广泛的

— set out　开发，出发，开始

— reform and opening-up strategy　改革开放战略

译文：我国通用航空的广泛商业应用并未正式启动，这是我国改革开放以来唯一没有开放的大产业。

12. For instance, Niagara Falls Park offers helicopter and hot-air balloon tours; every day in Colorado Canyon, dozens of helicopters take off one after another to render air cruise service to tourists; Helicopter sightseeing in Rotorua City of New Zealand tours across the Taravila Mountains to demonstrate the wonderful natural volcanic landscape of Rotorua; Brazil's Iguazu Falls, listed as a World Heritage Site by UNESCO can be appreciated with a panoramic view through helicopters; Tourists for Australian Great Barrier Reef, by the ride of helicopter, pass across the Coral Sea to dive, boat, and enjoy the wonderful experience.

— for instance　同于for example，例如

— hot-air ballon　热气球

— cruise service　巡游服务

— volcanic landscape　火山景色

— a panoramic view　全景

译文：如尼亚加拉瀑布公园开设了直升机和热气球的游览；科罗拉多大峡谷每天都有几十架直升机轮番起飞巡游服务游客；新西兰罗托鲁瓦城的直升机观光飞越塔拉威拉山，展示罗托鲁瓦神奇的自然火山景色；被联合国教科文组织列为世界自然遗产的巴西伊瓜苏瀑布通过直升机观光一览无余；澳大利亚大堡礁的游客乘坐直升机掠过珊瑚海去潜水、泛舟，享受奇妙感觉。

# Exercises

Section one: Answer the following questions according to the text.

1. What difficulties does China encounter during the development of general aviation?
2. Why is the low-altitude airspace essential to the development of GA?
3. What is low-altitude tourism and what are the advantages of low-altitude tourism?
4. Compared to China, what strategies has been done by some developed countries mentioned in the text?
5. What do you think of the opening up of the low-altitude airspace in China?

Section two: Match the words or phrases in column A with the definition in column B.

| A | B |
| --- | --- |
| 1. altitude | a. come to have or undergo a change of (physical features) |
| 2. develop | b. the formal act of approving |
| 3. destruction | c. an expanse of scenery that can be seen in a single view |
| 4. complex | d. close observation of a person or group (usually by the police) |
| 5. approval | e. elevation especially above sea level or above the earth's surface |
| 6. surveillance | f. a group of people living in a particular local area |
| 7. community | g. a conceptual whole made up of complicated and related parts |
| 8. landscape | h. the termination of something by causing so much damage |
| 9. reform | i. a line of argument rationalizing the course of action of a government |
| 10. policy | j. a change for the better as a result of correcting abuses |

Section three: Translate the following into Chinese.

1. Reform on low-altitude airspace management in China will be extended from 1000 meters to 3000 meters so as to realize gap-free connection between the

surveillance airspace and the report airspace.

2. The wide-scale commercialization of GA in China has also not been formally set out, the only large industry sector yet to be opened up since the country's reform and opening-up strategy.

3. China strongly encourages the development of tourism and sets to exert active efforts to promote low-altitude flight tourism.

4. Many countries have formulated long-term development strategies and corresponding policies, invested funds, trained relevant professionals, and established a sound law, regulation and standard system as well as GA industry organizations, which altogether put the whole international low-altitude tourism on the track of fostering a variety of industrial businesses with the core theme of low-altitude tourism.

5. Brazil's Iguazu Falls, listed as a World Heritage Site by UNESCO can be appreciated with a panoramic view through helicopters; Tourists for Australian Great Barrier Reef, by the ride of helicopter, pass across the Coral Sea to dive, boat, and enjoy the wonderful experience.

6. As always, the market has had great expectations for the opening-up of China's low-altitude airspace, and from the perspective of future reform tendency, it is already a common view among management teams to give place to the development of general aviation.

7. And the unavailability to operate flight and access to smooth airway turns out to be the key bottleneck in the development of general aviation in China.

8. Earlier, China's low-altitude airspace has been kept under control and its management reform lagged behind, which pose difficulties for GA planes in both taking off and landing.

### Section four: Translate the following into English.
1. 关于食品安全的争论引起了全国关注。
2. 那是一个私人企业的时代,众多小公司就是那时创建的。
3. 政府已经拒绝援助这家濒临破产的公司。
4. 有了新机器,我们终于有能力很好地完成该工作。
5. 1906年的大地震给旧金山大部分市区带来毁灭性的灾难。
6. 如果战争爆发,这个公司当然会采取行动保护自己在该国的经济利益。
7. 官员们将确保选举公正进行。
8. 她把经济衰退完全归咎于政府。
9. 我们当地的商店营业时间很合宜。
10. 政府消息人士估计火车票价将有一个涨幅达50%的长期上调。

## 1.3 GA airport
## 通用航空机场

### Background

Speaking of airports, many people will correlate general aviation airports with common civil aviation airports, assuming they are the same though they are not exactly the same. Civil aviation airport is the designated area for the take-off, landing, taxiing, parking and other activities of civil aviation aircraft, including its auxiliary buildings, plants and facilities. It can be grouped into general aviation airport and public transport airport and it excludes temporary airfield and special-purpose airport. General aviation airport certainly belongs to civil aviation airport, but undertakes more special flight missions such as business travel, air tourism, meteorological observation, agricultural and forestry spraying, fire alarm and patrol, air rescue and so on.

### Differences between civil aviation airport and general aviation airport

1. Different runway establishment: navigation facilities of runways in GA airport are usually primitive and generally not qualified for take-off and landing condition of large civil aviation aircrafts.

2. Different length of runway: the runway in civil aviation airport is fairly long whereas that in general aviation airport is relatively short.

3. Different physical looking of runway: in general aviation airport, runway is narrow and short or simply no runway but parking apron for helicopter to take off and land is established only. Runway in civil aviation airport is cement paved, but it could be a grassland, gravel area, snowfield, etc when it comes to general aviation airport.

4. Different runway location: civil aviation airport is all land-based. In case of mountain area, the top of the mountain may be leveled and in some cases the land may be reclaimed from coastal area so as to build the airport. General aviation airport, however, is much more flexible as it can be located on land, roof-based parking apron or even on the water.

5. Different kind of aircrafts parked in the airports: most of aircrafts undertaking GA flight operations are small, lightweight planes, helicopters, etc.; yet civil aviation airport is an area designated and specialized for large aircrafts to take off, land, taxi, park and engage in other activities.

6. Noted difference in application scenarios: civil aviation airport mainly assumes the passenger and cargo transportation. GA airport, instead, carries out flight

operations such as pilot training, aerial surveillance, meteorological observation, forest protection and agrochemical spraying, and also plays livelihood-related functions such as emergency response and rescue, business chartered flight, aerial photography, scenic spot sightseeing, air show, private flight, short-distance transport and air sports.

### Classification of general aviation airport

General aviation airport can be classified into type A and type B, depending on whether it's open to general public.

Type A general aviation airport is the GA airport open to the public, meaning that general public is allowed to enter into the airport to secure flight service or conduct their own flight activities;

Type B general aviation airport is the GA airport not open to the public, namely any other GA airport apart from type A general aviation airport.

**Classification table of GA airport**

| Category | | Definition |
|---|---|---|
| Type A | A1 class | Type A GA airport running commercial passenger-carrying flight activities with aircrafts configured with above 10 passenger seats. |
| | A2 class | Type A GA airport running commercial passenger-carrying flight activities with aircrafts configured with 5-9 passenger seats. |
| | A3 class | Other type A GA airport apart from A1 and A2 class airports. |
| Type B | | Refers to the GA airport not open to the public, namely any other GA airport apart from type A general aviation airport. |

Reference material: Regulation on the Classification of GA Airport (No.: MHF(2017)46)

As an industry driving greatly the growth of relevant businesses, general aviation contributes annually 0.6% of GDP to the United States, 90% of which were dependent on the contributions of mixed economies. An aviation project, after 10 years growth can benefit locally by input-output ratio of 1 : 80 and the job creation ratio of 1 : 12. In the United States, there're approximately above 2,000 GA airports engaging in general aviation businesses, 230,000 GA aircrafts, 19,000 general airports, 3,500 FBO, 10,000 MRO service providers, 61 AFSS and open airspaces, which make Americans fly in the sky as if they drove their private cars there. The GA industry in the country contributes above 150 billion USD worth of annual output and more than 1 million job openings. However, currently in China, the gap is huge as there're only about 300 general aviation airports, 300 GA enterprises and 2,000 registered GA aircrafts, which accounts for less than 1% of the total GA aircrafts across the world.

In conclusion, general aviation airport, being an important infrastructure and link in the GA industry, shall extend its business thought and operational practice. It should complement the large public air transport rather than alienate from it while running general aviation businesses, so as to truly realize the win-win growth of air transport and general aviation. Airport shall be defined accurately in terms of its public or non-public characteristic as per the lately published Regulation on the Classification of GA Airport. In case of public-featured GA airport which is open to the public, we shall understand its fundamental nature of being public infrastructure while constructing such airport, and shall not seek blindly for being big in size and complete in configurations, but on the basis of its practical operations, plan with proper size and under one master plan that will be valid all the way, grow to expand until such airport is upgraded into the transport airport at the right timing. In case of non-public GA airport which is not open to the public, we shall determine its construction size based on the investment size and developmental goal, and manage to yield satisfying returns on investment pursuant to the principle of starting it from easy to difficult and upgrading it step by step.

## General Vocabulary

correlate   *vt.* 使有相互关系；互相有关系
assume   *vt.* 假定；僭取；篡夺；夺取；擅用；侵占
designate   *vt.* 指定；指派；标出；把……定名为
taxi   *v.* (飞机) 滑行；乘出租车；用出租车送
aircraft   *n.* 飞机，航空器
auxiliary   *adj.* 辅助的；副的；附加的；(发动机、设备等) 备用的
facility   *n.* 设施；设备；容易；灵巧
public transport   公共交通；公共交通设施
exclude   *vt.* 排除；排斥；拒绝接纳；逐出
temporary   *adj.* 暂时的，临时的
undertake   *vt.* 承担，保证；从事；同意；试图
mission   *n.* 使命，任务；代表团；布道
meteorological   *adj.* 气象的；气象学的
spray   *n.* 喷雾，喷雾剂；喷雾器；水沫；*vt.* 喷射
fire alarm   火警；火警警报器
rescue   *v.* 营救，援救；(非正式) 防止……丢失；*n.* 营救，解救，援救；营救行动
kind   *n.* 种类；*adj.* 温和的；无害的
navigation   *n.* 航行；航海；导航
primitive   *adj.* 原始的，远古的；简单的，粗糙的
qualified   *adj.* 合格的；有资格的
land   *vt.* 使……登陆；使……陷于；将……卸下
length   *n.* 长度，长；时间的长短；(语) 音长

whereas　*conj.* 然而；鉴于
physical　*adj.* 物理的；身体的；物质的；根据自然规律的，符合自然法则的
narrow　*adj.* 狭窄的，有限的；勉强的；精密的；度量小的
apron　*n.* 围裙；停机坪；舞台口
cement　*n.* 水泥；接合剂；纽带；使人们团结的因素；黏固粉；牙骨质；沉积岩基质
pave　*v.* （用石板或砖）铺（路），铺砌，铺设；为某事物铺平道路，创造条件
grassland　*n.* 草原；牧草地
gravel　*n.* 碎石；砂砾
location　*n.* 位置；地点；外景拍摄场地
in case of　万一；如果发生；假设
level　*vt.* 使同等；对准；弄平
reclaim　*vt.* 开拓；回收再利用；改造某人，使某人悔改
coastal　*adj.* 沿海的；海岸的
roof　*n.* 屋顶；最高处，顶部；最高限度
lightweight　*adj.* 轻量的；给人印象不深的；无足轻重的；比通常重量轻的；浅薄的
specialized　*adj.* 专业的；专门的
scenario　*n.* 方案；情节；剧本；设想
assume　*vi.* 承担；假定；采取；呈现；装腔作势；多管闲事
carry out　*vt.* 执行，实行；贯彻；实现；完成
training　*n.* 训练；培养；瞄准；整枝
aerial　*adj.* 空中的，航空的；空气的；空想的
observation　*n.* 观察；监视；观察报告
protection　*n.* 保护；防卫；护照
agrochemical　*adj.* 农用化学品的
livelihood　*n.* 生计，生活；营生
response　*n.* 响应；反应；回答
photography　*n.* 摄影；摄影术
scenic　*adj.* 风景优美的；舞台的；戏剧的
sightsee　*vt.* 观光；游览
air show　飞行表演；空中表演
short-distance　短距离；短程
classification　*n.* 分类；类别，等级
depend on　取决于；依赖；依靠
enter　*vt.* 进入；开始；参加
obtain　*vi.* 获得；流行
conduct　*v.* 组织，实施，进行；指挥（音乐）；带领，引导；举止，表现；传导（热或电）
configure　*vt.* 安装；使成形
commercial　*adj.* 商业的；营利的；靠广告收入的
drive　*v.* 推动；钉；开凿；驱赶；迫使，逼迫；激励；挖掘；猛击（球）；吹，刮；猛

落；围赶（猎物）；使（抽象事物）发生或发展
  contribute   *vt.* 贡献，出力；投稿；捐献
  benefit   *vt.* 有益于，对……有益
  input-output ratio   投入产出比
  approximately   *adv.* 大约，近似地；近于
  FBO   *abbr.* 以固定机场为基地的飞行活动（fixed-base operation）
  MRO   *abbr.* 维修、修理服务（maintenance repair operating）
  AFSS   *abbr.* 自动飞行服务站（automated flight service station）
  register   *v.* 登记；（旅馆）登记住宿；挂号邮寄；表达（意见或情感）；显示（读数）
  in conclusion   总之；最后
  extend   *vt.* 延伸；扩大；推广；伸出；给予；使竭尽全力；对……估价
  operational   *adj.* 操作的；运作的
  complement   *vt.* 补足，补助
  alienate from   使疏远，离间；让与
  win-win   *adj.* 双赢的；互利互惠的
  characteristic   *n.* 特征；特性；特色
  fundamental   *adj.* 基本的，根本的
  blindly   *adv.* 盲目地；轻率地；摸索地
  configuration   *n.* 配置；结构；外形
  determine   *vt.* 决定，确定；判定，判决；限定
  yield   *v.* 出产（产品或作物）；产出（效果、收益等）；生息；屈服；放弃；停止争论；给（车辆）让路；（在外力、重压等下）屈曲
  pursuant   *adj.* 依据的；追赶的；随后的
  upgrade   *v.* 使（计算机、软件等）升级；改善（尤指服务）；给（飞机乘客或宾馆客人）升级；给（某人）升职；提高（某事物的）地位

## Notes

1. Civil aviation airport is the designated area for the take-off, landing, taxiing, parking and other activities of civil aviation aircraft, including its auxiliary buildings, plants and facilities. It can be grouped into general aviation airport and public transport airport and it excludes temporary airfield and special-purpose airport.
   - civil aviation airport   民用航空机场
   - group into   分成
   - public transport   公共运输
  译文：民用航空机场是指专供民用航空航空器起飞、降落、滑行、停放以及进行其他活动使用的划定区域，包括附属的建筑物、装置和设施。民用航空机场分为通用航空机场和公共运输机场；不包括临时机场和专用机场。

2. General aviation airport certainly belongs to civil aviation airport, but undertakes more special flight missions such as business travel, air tourism,

meteorological observation, agricultural and forestry spraying, fire alarm and patrol, air rescue and so on.

- belong to    属于，归于，附属于
- business travel    商务旅行，公务旅行
- air tourism    空中旅游
- air rescue    空中救援

译文：通用航空机场肯定属于民用航空机场，但承担了更多其他飞行任务，比如公务出差、空中旅游、气象探测、农林喷洒、消防警巡、空中救援等特殊飞行任务。

3. In general aviation airport, runway is narrow and short or simply no runway but parking apron for helicopter to take off and land is established only. Runway in civil aviation airport is cement paved, but it could be a grassland, gravel area, snowfield, etc when it comes to general aviation airport.

- parking apron    停机坪
- gravel area    砂石地
- snowfield    雪地

译文：通用航空机场跑道的长度宽度比较小，甚至没有跑道，只有供直升机起降的停机坪。民用航空机场的跑道是水泥跑道，但通航机场可以是草地、沙石地、雪地等。

4. In case of mountain area, the top of the mountain may be leveled and in some cases the land may be reclaimed from coastal area so as to build the airport. General aviation airport, however, is much more flexible as it can be located on land, roof-based parking apron or even on the water.

- mountain area    山区
- coastal area    沿海地区
- parking apron    停机坪

译文：山区需要削平山头，沿海填海造机场。但通用航空机场就灵活许多。可以在陆地上，也可以在楼顶停机坪，甚至在水里起飞。

5. Yet civil aviation airport is an area designated and specialized for large aircrafts to take off, land, taxi, park and engage in other activities.

- large aircraft    大型航空器
- engage in    参与，参加，从事于

译文：民用航空机场则是专供大型航空器起飞、降落、滑行、停放以及进行其他活动使用的划定区域。

6. GA airport, instead, carries out flight operations such as pilot training, aerial surveillance, meteorological observation, forest protection and agrochemical spraying, and also plays livelihood-related functions such as emergency response and rescue, business chartered flight, aerial photography, scenic spot sightseeing, air show, private flight, short-distance transport and air sports.

- carry out    执行，实行；贯彻；实现；完成
- pilot training    飞行员训练
- agrochemical spraying    喷洒农药

— livelihood-related  民生相关的
— aerial photography  航空摄影
— air show  空中表演

译文：通用航空机场则开展飞行员培训、空中巡查、空中测绘、气象探测、防林护林、喷洒农药等作业飞行，以及应急救援、商务包机、空中摄影、景点观光、空中表演、私人飞行、短途运输、航空运动等民生功能。

7. Type A GA airport running commercial passenger-carrying flight activities with aircrafts configured with above 10 passenger seats.

— commercial passenger-carrying  商业载客
— configure with  配置，配备

译文：含有使用乘客座位数在10座以上的航空器开展商业载客飞行活动的A类通用机场。

8. As an industry driving greatly the growth of relevant businesses, general aviation contributes annually 0.6% of GDP to the United States, 90% of which were dependent on the contributions of mixed economies. An aviation project, after 10 years growth can benefit locally by input-output ratio of 1∶80 and the job creation ratio of 1∶12.

— contribute to  有助于；捐献；带来，促成
— the United States  美国
— dependent on  依赖于；依靠
— mixed economies  混合经济，混合经济是各种不同因素在一定社会制度下混合的经济运行体制
— input-output ratio  投入产出比，是指项目全部投资与运行寿命期内产出的工业增加值总和之比。它是适用于科技项目、技术改造项目和设备更新项目的经济效果评价指标。其值越大，表明经济效果越好

译文：作为带动性强的行业，通用航空每年为美国贡献0.6%的GDP，其中90%都依赖综合经济带动。一个航空项目发展10年后给当地带来的效益，投入产出比可达1∶80，通用航空的就业带动比是1∶12。

9. In conclusion, general aviation airport, being an important infrastructure and link in the GA industry, shall extend its business thought and operational practice. It should complement the large public air transport rather than alienate from it while running general aviation businesses, so as to truly realize the win-win growth of air transport and general aviation.

— in conclusion  总之；最后
— infrastructure  基础设施；公共建设；下部构造
— rather than  而不是；宁可……也不愿
— win-win  双赢的；互利互惠的

译文：综上所述，通用航空机场作为通用航空业的重要基础设施和节点，要扩展经营思路和经营方式，在开展通用航空业务的同时，做好大型公共航空运输的补充而不是与其割裂，真正实现运输航空和通用航空两翼齐飞。

10. Airport shall be defined accurately in terms of its public or non-public

characteristic as per the lately published Regulation on the Classification of GA Airport.

— in terms of　依据；按照；在……方面；以……措辞

— Regulation on the Classification of GA Airport《通用机场分类管理办法》(以下简称《办法》)是为贯彻落实国务院办公厅《关于促进通用航空业发展的指导意见》要求，促进通用机场健康发展制定，由民航局于2017年4月14日印发并实施。《办法》按照通用机场是否对公众开放分为A、B两类。A类为对公众开放的通用机场，允许公众进入以获取飞行服务或自行开展飞行活动；B类则为不对公众开放的通用机场。另外，基于其对公众利益的影响程度，《办法》又将A类通用机场分为三级。其中，含有使用乘客座位数在10座以上的航空器开展商业载客飞行活动的为A1级通用机场，使用座位数在5~9座之间的航空器开展商业载客飞行活动的为A2级通用机场，其余均为A3级通用机场。今后，我国所有通用机场的建设与运行管理都将遵循此《办法》。

译文：根据最新颁布的《通用机场分类管理办法》，准确界定机场的公共性和非公共性定位。

11. In case of non-public GA airport which is not open to the public, we shall determine its construction size based on the investment size and developmental goal, and manage to yield satisfying returns on investment pursuant to the principle of starting it from easy to difficult and upgrading it step by step.

— based on　基于，以……为根据；在……基础上
— pursuant to　依据，依照，按照
— step by step　逐步地，逐渐地

译文：对不对公众开放的非公共性通用机场，依据投资实力和发展目标来确定其建设规模，可遵循"从易到难，逐步升级"的原则，实现良好的投资收益。

## Exercises

Section one: Answer the following questions according to the text.
1. What do we know about Regulation on the Classification of GA Airport?
2. How do we classify civil aviation airport and general aviation airport?
3. What kinds of flight missions does general aviation airport undertake?
4. What are major differences between civil aviation airport and general aviation airport?
5. Compared to China, what is the current situation of American general aviation market?

Section two: Match the words or phrases in column A with the definition in column B.

| A | B |
| --- | --- |
| 1. navigation | a. people in general considered as a whole |
| 2. configure | b. a contract to hire or lease transportation |

3. public　　　　　　　c. the act of making and recording a measurement
4. charter　　　　　　 d. able to adjust readily to different conditions
5. pilot　　　　　　　 e. the property created by the space between two objects or points
6. observation　　　　 f. the guidance of ships or airplanes from place to place
7. distance　　　　　　g. set up for a particular purpose
8. operation　　　　　 h. someone who is licensed to operate an aircraft in flight
9. flexible　　　　　　 i. a business especially one run on a large scale
10. civil　　　　　　　 j. applying to ordinary citizens

## Section three: Translate the following into Chinese.

1. Civil aviation airport is the designated area for the take-off, landing, taxiing, parking and other activities of civil aviation aircraft, including its auxiliary buildings, plants and facilities. It can be grouped into general aviation airport and public transport airport and it excludes temporary airfield and special-purpose airport.

2. In general aviation airport, runway is narrow and short or simply no runway but parking apron for helicopter to take off and land is established only.

3. GA airport, instead, carries out flight operations such as pilot training, aerial surveillance, meteorological observation, forest protection and agrochemical spraying, and also plays livelihood-related functions such as emergency response and rescue, business chartered flight, aerial photography, scenic spot sightseeing, air show, private flight, short-distance transport and air sports.

4. Type A general aviation airport is the GA airport open to the public, meaning that general public is allowed to enter into the airport to obtain flight service or conduct their own flight activities.

5. An aviation project, after 10 years growth can benefit locally by input-output ratio of 1∶80 and the job creation ratio of 1∶12. In the United States, there're approximately above 2,000 GA airports engaging in general aviation businesses, 230,000 GA aircrafts, 19,000 general airports, 3,500 FBO, 10,000 MRO service providers, 61 AFSS and open airspaces, which make Americans fly in the sky as if they drove their private cars there.

6. In case of public-featured GA airport which is open to the public, we shall understand its fundamental nature of being public infrastructure while constructing such airport, and shall not seek blindly for being big in size and complete in configurations, but on the basis of its practical operations, plan with proper size and under one master plan that will be valid all the way, grow to expand until such airport is upgraded into the transport airport at the right timing.

7. Airport shall be defined accurately in terms of its public or non-public characteristic as per the lately published Regulation on the Classification of GA Airport.

## Section four: Translate the following into English.

1. 最初估计申请选修这门课程的学生也就是50到100人，可实际上报名的有120人。
2. 我的时间安排很灵活——下周任何一天我都可以安排与你见面。
3. 20世纪80年代他们扩大了零售业务。
4. 这座城市的交通系统运行效率很高，在全欧洲名列前茅。
5. 我们需要的人必须要高效干练，善于组织筹划，能使这个办事处顺畅运作。
6. 这种药物治疗头痛极其有效。
7. 地基必须进行加固以防房屋继续下陷。
8. 她的态度一直在改进。
9. 证人在法庭上的证词和他对警方的陈述不一致。
10. 考虑到她只到这里来过一次，能找到路已经很不错了。

# 1.4
# GA in China
## 中国通用航空

## Background

The general aviation industry began in China in 1951 and was initially established under harsh and poor historical conditions in which the older generation pioneers the way to make contributions to the development of GA in new China. With the entrance of 21st century, general aviation begins to flourish, presenting new dynamism and potential for development.

（1）Operational Capabilities

1) Total number of GA aircrafts

By the end of 2006, there's a sum of 707 registered airplanes and helicopters (including planes and helicopters for training and flight inspection purpose) in a total of 46 models across the whole GA industries in China, among which 575 aircrafts are airplanes and 132 helicopters, covering respectively 22 and 24 models. The total aircrafts in 2006 was 92 more in number from a year earlier, and the total fleets, airplanes and helicopters saw a respective net increase of 36.6%, 27.3%, 55.2% over the year of 1996. Thanks to the renewal of certain old transport capacity, equipments are more advanced and their safety levels improved, and in general, the GA's capacity in both air transport and actual production have boosted greatly than that a decade ago.

Nevertheless, the achievement is still way far away from the expectations set forth in the 11th Five-year Plan.

2) Utilization rate of GA aircrafts

Statistically the GA aircrafts in China are not highly utilized. In 2002, their overall average annual production hours were 219 and average daily production hour was less than 1 hour with the most employed aircrafts respectively being Super Puma, EC155B, Caravan, CL-604, Z-11, S-76, A109 and Hawker 800XP.

3) Airport and operational staff

As at the end of 2005, there were 68 general aviation airports and 329 temporary airports (taking-off and landing airfield). And by the end of 2006, pilots in possession of valid pilot license across the whole lines of Chinese civil aviation business totalled 16031 persons, among which 6901 persons held airline transport pilot license, 7757 persons commercial pilot license and the rest of 1373 persons private pilot license.

## (2) Productive operations

Since 1951, the beginning year of general aviation in new China, GA had turned out to be so widely popular among different sectors that it has gained a rapid development. In 1980, the GA's flight operation hours was 42700h, but since then, the yearly operation hours had been fluctuating around 40000h until the year of 2000 after which the GA welcomed another sound run-up. In 2006, the GA's flight operations hours rose to a record high of 98700h, a 16% increase over 2005, and the total GA flight hours were 180000h in the same year, according to the statistics of international civil aviation organization. Over the period of 9th and 10th Five-year Plan, namely from 1996 to 2005, the GA flight operations grew annually by 9.1% on average. Accumulatively the completed GA flight operation hours during the period of 10th Five-year Plan alone were 336000h, an increase of 59% over the 9th Five-year Plan period or an annual growth rate of about 11% within the five years, among which the airborne operational flights accounted for 46.6% and the teaching & training flights for 53.4%.

## (3) Market scale of GA industry

Initially Chinese general aviation, with its operation price set by government and paid by fiscal funds, serves mainly for government departments or state-owned enterprises. And as the GA's operational projects and fields develop, general aviation operations from the mid-1950s to the mid-1970s are mainly applied in some projects which were difficult to meet the time and technical requirements by conventional means in the process of industrial and agricultural production and construction, such as agroforestry flight, aerial forest fire-protection, aerial photography and aeromagnetic survey. From the late 1970s to the mid-1980s, however, the priority of GA operations gave way to meet the needs of energy exploration and development in the national economic development. The most representative projects at the time were offshore and onshore petroleum services. And from the end of 1980s to the middle of 1990s, the majority of GA operations were still agroforestry flight and helicopter services for energy exploration and development, whereas general aviation had tended to open its service to the society and the public as most represented projects including air tourism, short-distance transportation and business flight.

In recent years, with the continuous and rapid national economic growth and the steady progression of marketization, GA's market has been more significantly restructured. The operation projects of general aviation have been expanded to more than 100 types over 7 major fields from the two or three types, covering main business such as aerial photography, aerial prospecting, artificial precipitation, aerial forest fire-protection, locust killing by aircraft, agroforestry flight, etc, and serving more than ten departments and industries including those for national economic construction, scientific research and social development across all the nationwide provinces, autonomous regions and municipalities except for Taiwan and Tibet.

At present, the market structure of general aviation shows that the traditional operation projects still dominates, but their shares have been falling, and meanwhile the market's self-driven demands are growing year after year, for which the operation services are priced completely based on market supply and demand, and such operation services with good momentum of growth include petroleum service, business flight, medical rescue and flight training etc. Moreover, as the state-owned enterprises and institutions are buying more and more their own business flights (including helicopters), the demand for passenger ferry service and medical rescue by helicopters and for the shuttle flight between short-distant cities by small airplanes are booming. When the national economy grows to certain level, the market for GA will scale up, and the market will also be more able to consume the services.

In 2003, Civil Aviation Administration of China delegated the right to approve the establishment of general aviation company to the regional administration authority of civil aviation to create a relaxed environment for the development of GA. And up to the end of 2006, there were 69 enterprises engaged in GA-related business activities, 30 approved enterprises preparing to establish GA and nearly 8000 people employed in the industry across the board, which turned out to be a great improvement in comparison with the situation where there were only 29 GA operational companies and about 4000 employees in the trade before 1995 when GA industry suffered severe cutback.

On the basis of the original form of principal for GA activities, entity and individual engaging in non-operational general aviation activities are also incorporated. And nowadays there're 12 entities and 2 natural persons registered for engaging in non-operational GA activities.

Among these GA companies, state-owned ones and those key ones play a leading role in seeing that services are available for national mandatory rescue and disaster relief and other major GA missions. In 2006, the market size of general aviation was worth 1.79 billion yuan, of which about 700 million yuan was honored by governments at all levels.

Though the general aviation has made certain progress in light of its market and revenue situation, its share out of the total Chinese civil aviation industry remains small. To say the least, the income of general aviation makes up less than 1% of that of the whole civil aviation.

Among the current range of Chinese GA services, the operation on offshore oil drilling and exploration relatively stands out (followed by the agriculture and forestry operations), thanks mainly to the country's opening-up to the domestic offshore oil development. And as China covers a vast territory and has different region-specific traditional customs, economic basis, resources as well as development speed, the general aviation also features noticeable regional differences. For example, the agricultural and forestry aviations are better developed in the northeast and northwest China, and the offshore oil services are more advanced in the coastal areas of east and south China. Also, in years say 2001 and 2002, the total industrial aviation operations were 21752h, out of which 46.4% was contributed by middle south China, and the total agricultural aviation operations were recorded at 30754h, northeast China taking up 41.2%.

The regions on a good track of developing general aviation are Guangdong (mainly offshore petroleum services), Xinjiang (onshore petroleum and agricultural services) and Heilongjiang (agriculture, forestry and animal husbandry).

## General Vocabulary

initially   *adv.* 最初，首先；开头
harsh   *adj.* 严厉的；严酷的；刺耳的；粗糙的；刺目的；丑陋的
historical   *adj.* 历史的；史学的；基于史实的
pioneer   *n.* 先锋；拓荒者
development   *n.* 发展；开发；发育；住宅小区（专指由同一开发商开发的）；显影
entrance   *n.* 入口；进入
flourish   *v.* 繁荣，茂盛；茁壮成长，处于旺盛时期；挥舞；炫耀
dynamism   *n.* 活力；动态；物力论；推动力；精神动力作用
inspection   *n.* 视察，检查
model   *n.* 模型；典型；模范；模特儿；样式
respectively   *adv.* 分别地；各自地，独自地
thanks to   由于，幸亏
renewal   *n.* 更新，恢复；复兴；补充；革新；续借；重申
capacity   *n.* 能力；容量；资格，地位；生产力
equipment   *n.* 设备，装备；器材
advanced   *adj.* 先进的；高级的；晚期的；年老的
safety   *n.* 安全；保险；安全设备；保险装置；安打

boost　vt. 促进；增加；支援
achievement　n. 成就；完成；达到；成绩
utilization　n. 利用，使用
statistically　adv. 统计地；统计学上
daily　adj. 日常的；每日的
possession　n. 拥有；财产；领地；自制；着迷
license　n. 执照，许可证；特许
private pilot license　私人飞行员执照
beginning　n. 开始；起点
Five-year Plan　五年计划
fluctuate　vi. 波动；涨落；动摇
run-up　n. 助跑；预备阶段；抬高；急剧增长
statistics　n. 统计；统计学；统计资料
organization　n. 组织；机构；体制；团体
annually　adv. 每年；一年一次
on average　平均；普通，通常
accumulatively　adv. 累积地
airborne　adj. 空运的；空气传播的；风媒的
scale　n. 规模；比例；鳞；刻度；天平；数值范围
fiscal　adj. 会计的，财政的；国库的
fund　n. 基金；资金；存款
state-owned　adj. 国有的；国营的；州立的
technical　adj. 工艺的，科技的；技术上的；专门的
requirement　n. 要求；必要条件；必需品
conventional　adj. 符合习俗的，传统的；常见的；惯例的
agricultural　adj. 农业的；农艺的
agroforestry　n. 农用林业；农林复合经营；经济林
fire-protection　消防；消防处
aeromagnetic　adj. 航空磁测的；空中探测地磁的
priority　n. 优先；优先权；优先次序；优先考虑的事
exploration　n. 探测；探究；踏勘
national　adj. 国家的；国民的；民族的；国立的
majority　n. 多数；成年
tourism　n. 旅游业；游览
steady　adj. 稳定的；不变的；沉着的
progression　n. 前进；连续
marketization　n. 市场化；自由市场经济化；向自由市场经济转化
significantly　adv. 显著地；相当数量地
restructure　v. 改组；重组（困难企业的债务）
prospect　vi. 勘探，找矿

# Unit 1　General introduction of GA 通用航空概述

artificial　*adj.* 人造的；仿造的；虚伪的；非原产地的；武断的
precipitation　*n.* 沉淀，沉淀物；降水；冰雹；坠落；鲁莽
locust　*n.* 蝗虫，蚱蜢
research　*n.* 研究；调查
nationwide　*adj.* 全国范围的；全国性的
province　*n.* 省；领域；职权
autonomous　*adj.* 自治的；自主的；自发的
region　*n.* 地区；范围；部位
municipality　*n.* 市政当局；自治市或区
except for　除了……以外；要不是由于
Tibet　*n.* 西藏
traditional　*adj.* 传统的；惯例的
dominate　*vt.* 控制；支配；占优势；在……中占主要地位
share　*n.* 份，份额；股份；责任，贡献
meanwhile　*adv.* 同时，其间
self-driven　*adj.* 自行驱动的；自励的
momentum　*n.* 势头；动量；动力；冲力
petroleum　*n.* 石油
ferry　*n.* 渡船；摆渡；渡口
shuttle　*n.* 航天飞机；穿梭；梭子；穿梭班机、公共汽车等
consume　*vt.* 消耗，消费；使……着迷；挥霍
delegate　*v.* 授（权），把……委托给他人；委派……为代表，任命
authority　*n.* 权威；权力；当局
prepare　*vt.* 准备；使适合；装备；起草
cutback　*n.* 减少，削减；情节倒叙
incorporate　*vt.* 包含，吸收；体现；把……合并
leading　*adj.* 领导的；主要的
mandatory　*adj.* 强制的；托管的；命令的
honor　*vt.* 尊敬（与honour同）；给……以荣誉
in light of　根据；鉴于；从……观点
revenue　*n.* 税收收入；财政收入；收益
make up　组成；补足；化妆；编造
drill　*vi.* 钻孔；训练
stand out　突出；站出来；坚持到底；坚决反对
territory　*n.* 领土，领域；范围；地域；版图
custom　*n.* 习惯，惯例；风俗；海关，关税；经常光顾；[总称]（经常性的）顾客
feature　*vt.* 特写；以……为特色；由……主演
noticeable　*adj.* 显而易见的，显著的；值得注意的
husbandry　*n.* 饲养；务农，耕种；家政

## Notes

1. The general aviation industry began in China in 1951 and was initially established under harsh and poor historical conditions in which the older generation pioneers the way to make contributions to the development of GA in new China.

- historical condition  历史条件下
- make contributions to  为……作贡献

译文：中国通用航空事业始于1951年。通用航空组建初期，老一辈的开拓者在"一穷二白"的历史条件下，为新中国建设做出了贡献。

2. Thanks to the renewal of certain old transport capacity, equipments are more advanced and their safety levels improved, and in general, the GA's capacity in both air transport and actual production have boosted greatly than that a decade ago. Nevertheless, the achievement is still way far away from the expectations set forth in the 11th Five-year Plan.

- transport capacity  运输能力，输送能力
- far away from  远离；离……远

译文：由于部分老旧运力得到更新，设备更加先进，安全水平得到改善，通用航空运力总体性和实际生产能力比10年前有较大的提高，但是，与"十一五"规划的要求还相距甚远。

3. And by the end of 2006, pilots in possession of valid pilot license across the whole lines of Chinese civil aviation business totalled 16031 persons, among which 6901 persons held airline transport pilot license, 7757 persons commercial pilot license and the rest of 1373 persons private pilot license.

- by the end of  到……结束时；到……时为止；在……之前
- transport pilot license  运输飞行员执照
- commercial pilot license  商业飞行员执照
- private pilot license  私人飞行员执照

译文：截至2006年年底，我国民航全行业持有有效飞行驾驶员执照的飞行人员共16031人，其中持有航线运输驾驶员执照的6901人，商用驾驶员执照的7757人，私用驾驶员执照的1373人。

4. In 1980, the GA's flight operation hours was 42700h, but since then, the yearly operation hours had been fluctuating around 40000h until the year of 2000 after which the GA welcomed another sound run-up.

- run-up  助跑；预备阶段；抬高；急剧增长

译文：1980年通用航空飞行作业为42700h，但从1980年一直到2000年，年作业量徘徊在40000h左右，进入2000年以后，通用航空有了较大的发展。

5. Over the period of 9th and 10th Five-year Plan, namely from 1996 to 2005, the GA flight operations grew annually by 9.1% on average. Accumulatively the completed GA flight operation hours during the period of 10th Five-year Plan alone were 336000h, an increase of 59% over the 9th Five-year Plan period or an annual growth rate of about 11% within the five years, among which the airborne operational flights accounted for 46.6%

and the teaching & training flights for 53.4%.

- the 9th Five-year Plan  "九五"规划
- the 10th Five-year Plan  "十五"规划
- training flights  飞行训练

译文：1996—2005年的年平均增长速度为9.1%。我国"十五"期间累计完成通用航空作业飞行336000h，比"九五"期间增长59%，五年平均增长率为11%左右，其中，空中作业飞行占46.6%，教学和训练飞行占53.4%。

6. Initially Chinese general aviation, with its operation price set by government and paid by fiscal funds, serves mainly for government departments or state-owned enterprises.

- fiscal fund  财政资金
- government department  政府部门
- state-owned enterprise  国营企业，国有企业

译文：早期中国通用航空市场的主要服务对象为政府部门或国有企业，作业价格依靠政府制定的作业收费标准执行，作业费用由财政支付。

7. And as the GA's operational projects and fields develop, general aviation operations from the mid-1950s to the mid-1970s are mainly applied in some projects which are difficult to meet the time and technical requirements by conventional means in the process of industrial and agricultural production and construction, such as agroforestry flight, aerial forest fire-protection, aerial photography and aeromagnetic survey.

- technical requirement  技术要求；技术条件；技巧请求
- agricultural production  农业生产
- agroforestry flight  农林化飞行
- aeromagnetic survey  航磁测量

译文：从作业项目的发展看，20世纪50年代中期至70年代中期，通用航空作业多集中在工农业生产建设中一些靠常规手段难以达到时间和工作技术要求的项目，如农林化飞行、航空护林、航空摄影和航空磁法测量等。

8. And from the end of 1980s to the middle of 1990s, the majority of GA operations were still agroforestry flight and helicopter services for energy exploration and development, whereas general aviation had tended to open its service to the society and the public as most represented projects including air tourism, short-distance transportation and business flight.

- energy exploration  能源勘探
- tend to  趋向；注意；易于；有……的倾向
- short-distance transportation  短途运输
- business flight  公务飞行

译文：80年代末至90年代中期，占通用航空作业比重大部分的仍是农林化飞行，能源勘探和开发中的直升机服务，但通用航空已出现向社会和公众提供服务的倾向，其最有代表的项目为空中游览、短途运输和公务飞行。

9. The operation projects of general aviation have been expanded to more than 100 types over 7 major fields from the two or three types, covering main business such as aerial photography, aerial prospecting, artificial precipitation, aerial forest fire-protection, locust killing by aircraft, agroforestry flight, etc, and serving more than ten departments and industries including those for national economic construction, scientific research and social development across all the nationwide provinces, autonomous regions and municipalities except for Taiwan and Tibet.

- aerial photography　航空摄影
- aerial prospecting　航空探矿
- artificial precipitation　人工降水
- scientific research　科学研究
- autonomous region　自治区

译文：通用航空的作业项目已由两三种增加到7大类100多种，主要经营项目包括航空摄影、航空探矿、人工降水、航空护林、飞机灭蝗、农林化飞行等，服务领域涉及国民经济建设、科学研究和社会发展等10多个部门和行业，作业范围遍及全国除台湾和西藏外的所有省、自治区和直辖市。

10. At present, the market structure of general aviation shows that the traditional operation projects still dominates, but their shares have been falling, and meanwhile the market's self-driven demands are growing year after year, for which the operation services are priced completely based on market supply and demand, and such operation services with good momentum of growth include petroleum service, business flight, medical rescue and flight training etc.

- self-driven　自行驱动的；自励的
- market supply　市场供应
- medical rescue　医疗救援，医疗救助
- flight training　飞行训练
- etc　等等，及其他 (et cetera)

译文：目前通用航空的市场结构表现为，传统作业项目仍占相当比例，但份额已在逐步下降；自发的市场需求在逐年增长，其作业价格完全按照市场供需关系来确定，发展势头较好的经营项目有石油服务、公务飞行、医疗救护、培训飞行等。

11. In 2003, Civil Aviation Administration of China delegated the right to approve the establishment of general aviation company to the regional administration authority of civil aviation to create a relaxed environment for the development of GA.

- Civil Aviation Administration of China　中国民用航空局（简称：中国民航局或民航局，英文缩写CAAC）是中华人民共和国国务院主管民用航空事业的由部委管理的国家局，归交通运输部管理。

译文：2003年，中国民航总局把成立通用航空公司的经营许可审批权下放到民航地区管理局，给通用航空的发展创造了一个宽松的环境。

12. Among these GA companies, state-owned ones and those key ones play a leading role in seeing that services are available for national mandatory rescue and

disaster relief and other major GA missions.

- state-owned　国有，州立的，国营的，国有企业
- play a leading role　起着主导的作用，发挥主导作用
- disaster relief　赈灾；灾难援助

译文：在这些通用航空公司中，国有通用航空企业和骨干通用航空企业起着主导作用，它们为保证国家指令性抢险救灾和重大通用航空任务提供服务。

13. Among the current range of Chinese GA services, the operation on offshore oil drilling and exploration relatively stands out (followed by the agriculture and forestry operations), thanks mainly to the country's opening-up to the domestic offshore oil development. And as China covers a vast territory and has different region-specific traditional customs, economic basis, resources as well as development speed, the general aviation also features noticeable regional difference.

- oil drilling　石油，钻井探油，石油钻探
- stand out　突出；站出来；坚持到底；坚决反对
- opening-up　开放
- region-specific　区域特异性的

译文：在目前我国的通用航空服务业中，海洋石油开采和勘探是运营较成功的领域，主要得益于国家对国内近海石油开发的开放；其次是农林作业。我国地域辽阔，不同地区的传统习俗不同、经济基础不同、资源不同，发展速度也不同，由此使通用航空的地区性差异十分明显。

## Exercises

### Section one: Answer the following questions according to the text.

1. In respective of operational capabilities, how did the total number of GA aircrafts in China change from 2005 to 2006?
2. What are the Chinese most employed aircrafts in 2002?
3. How did the GA's flight operation hours change from 1980 to 2006?
4. What were the most representative projects from the late 1970s to the mid-1980s?
5. Why does Chinese general aviation have noticeable regional differences?

### Section two: Match the words or phrases in column A with the definition in column B.

| A | B |
| --- | --- |
| 1. offshore | a. comparatively late in a course of development |
| 2. territory | b. a person appointed or elected to represent others |
| 3. advanced | c. a traveler riding in a vehicle who is not operating it |
| 4. register | d. increasing by successive addition |
| 5. delegate | e. be unstable |

| | |
|---|---|
| 6. passenger | f. at some distance from the shore |
| 7. photography | g. a region marked off for administrative or other purposes |
| 8. accumulative | h. an official written record of names or events or transactions |
| 9. fluctuate | i. a legal document giving official permission to do something |
| 10. license | j. the act of taking and printing photographs |

### Section three: Translate the following into Chinese.

1. In 2006, the GA's flight operations hours rised to a record high of 98700h, a 16% increase over 2005, and the total GA flight hours were 180000h in the same year, according to the statistics of international civil aviation organization.

2. From the late 1970s to the mid-1980s, however, the priority of GA operations gives way to meet the needs of energy exploration and development in the national economic development. The most representative projects at the time are offshore and onshore petroleum services.

3. Moreover, as the state-owned enterprises and institutions are buying more and more their own business flights (including helicopters), the demand for passenger ferry service and medical rescue by helicopters and for the shuttle flight between short-distant cities by small airplanes are booming.

4. And up to the end of 2006, there're 69 enterprises engaged in GA-related business activities, 30 approved enterprises preparing to establish GA and nearly 8000 people employed in the industry across the board, which turn out to be a great improvement in comparison with the situation where there're only 29 GA operational companies and about 4000 employees in the trade before 1995 when GA industry suffers severe cutback.

5. On the basis of the original form of principal for GA activities, entity and individual engaging in non-operational general aviation activities are also incorporated.

6. Though the general aviation has made certain progress in light of its market and revenue situation, its share out of the total Chinese civil aviation industry remains small.

7. The regions on a good track of developing general aviation are Guangdong (mainly offshore petroleum services), Xinjiang (onshore petroleum and agricultural services) and Heilongjiang (agriculture, forestry and animal husbandry).

### Section four: Translate the following into English.
1. 那场火灾给教堂造成了重大损失。
2. 为了省电，我们在减少中央供热。
3. 非常有必要与员工、家长、学生以及其他各方建立有效的沟通渠道。

4. 三个月后我们在新的工作中站稳了脚跟。
5. 要保证实验的有效性，就必须精确地记录数据。
6. 这本小册子会告诉你飓风到来前如何做好准备的基本知识。
7. 批评只是削弱了他们的信心。
8. 在羞涩的外表下，她其实是个非常热心的女孩。
9. 教育应和孩子的需要挂钩。
10. 欲知详情，请参阅有关单页广告。

# Unit 2

# General Aviation in USA
# 美国通用航空

# 2.1 Aviation town in the United States
# 美国航空小镇

## Background

Aviation town, an exotic concept, means a cluster of urbanized communities which is themed by the core business and infrastructure of general aviation, and holds a variety of functions such as manufacturing, residence, business, recreation, tourism and exhibition. Underpinned by the full-fledged GA industry and its rich GA culture, aviation town creates a lifestyle that is compatible with the development of general aviation. And due to different geographical locations, and cultural and industrial background, aviation towns in foreign countries have been booming with distinctive features, which can be generalized into three types, i.e. industry-oriented, environment-oriented and activity-oriented. The industry-oriented town attracts people by its industry and it includes commercial headquarter-based town and town of manufacturing base. Environment-oriented town draws people's affection by its environment and it can be grouped into tourism-and-vacation-specific one and livable-and-experiencing-specific one. Activity-oriented town gathers people by holding activities such as exhibitions, amusements and events.

Now there're approximately more than 600 aviation towns in the world, and nearly 500 of which are located in the USA. Of those 500 towns, over 300 residential airparks are included, and Spruce Creek is the largest and most versatile one. The Spruce Creek Airport was established from a navy airfield which was constructed in 1943. The airfield was then taken over by local governments after World war II, and was sold to private owner in the 1970s for commercialization. Today in the Spruce Creek Airport, there're a 1,214-meter-long runway, two FBOs, more than 600 villas furnished with hangars and connected to airport runway, one championship-graded golf course and several fancy restaurants and clubs. And the whole community is isolated from outside world by enclosure and also falls under 24-hour patrol, with primarily upper-class wealthy people and celebrities living inside. A 700m$^2$ villa, if connected directly to the runway and equipped with both hangar and swimming pool, could cost more than 2.5 million US dollars.

In the United States, there're also many fly-in communities at the service of aviation fans, and globally the rest of aviation towns are located in Europe, Australia, South America and so on. Originally from the United States, the aviation town has

seen its emergence and further development in the country, thanks to the country's adequate airport resources and large number of pilots. After World War II, multitudes of ex-service pilots gathered in many abandoned military airports, and over time, many fly-in communities surrounding such airports, or known as residence-oriented aviation town took shape. Meanwhile, America's GA industry has made a giant step forward. Based on the enormous consuming market, the manufacturing industry of general aviation in the USA attained unprecedented development after World War II, and gradually a town came into being, which was underpinned by the aviation manufacture and correlated industries. Compared with the residence-oriented aviation town, such industry-oriented town is dyed with strong industrial characteristics. As to eco-tour town, it is mainly distributed around the famous tourism attractions all over the world.

General aviation has been integrated into American everyday life. The mature GA industry and its mass population have made flying a lifestyle, and the vast expanse of lands, together with the specific living habits, shapes the form of towns and promotes their development. Among them, Florida enjoys the largest number of such aviation towns in the USA, i.e. about 70. As to GA airports in the United States, they have a wide range of property rights, including private, national, state, municipal and

community ownership, and most of which are owned by governments at all levels though. As airport by nature is a non-profit unit, its operating income can only be allocated to offset overheads and personnel payrolls, and any of surplus will be deposited in a special accounts under the supervision of municipal council, which then strictly confines the airport to be a public infrastructure.

With the progression of general aviation and the startup of private flight, the idea of aviation town has become more and more aware in China, and as a matter of fact, many cities plan to build or are building aviation towns. For the moment, mass construction of aviation towns in American way is still far short of basic conditions in China, hence the majority of GA town currently is operated for tourism and commercial purpose. Nevertheless, there're nearly 200 regions claiming to have the aviation town in the pipeline, which lays a good foundation for the startup and advancement of aviation town in China. The aviation town extends the whole industrial chain of low-altitude tourism, forming a comprehensive tourism destination supported by a variety of businesses, including sightseeing, flight training, real estate development, extension of theme entertainments, and aviation festivals and events.

Different from the aviation towns abroad, Chinese ones are devoid of specific developing orientation as well as clear development methods. Available airport, flight and airspace are the basic preconditions for the development of GA-featured town. And in China, most of aviation towns are planned and constructed surrounding the established GA airports, and some are constructed as aviation industry complex buildings which are based on the original aviation manufacturing industries and facilitated with relevant ancillary infrastructure. At present, only a small part of aviation towns have begun to take shape and the remaining are still in plan or under construction, for which the reasons are the weak legal-person-based GA manufacturing industry, restricted low-altitude airspace, imperfectly configured GA airports and flight-related infrastructure, as well as the lack of relevant supporting service and user basis for private flight.

Varied property rights of GA airports in the USA (including private, national, state, municipal and community ownership) does not bother the government as what matters is that whether the airport is public-purpose or private-purpose. And as far as the airport can satisfy the test of enough openness to the public and the low charges for public welfare consideration, such airport may apply to FAA for AIP, a subsidy as high as 95% for the facilitation of its reconstruction and extension. Except for the aforementioned small-scale GA airport which serves only say one community, most of the GA airports are owned by American government at all levels. GA airport itself being a non-profit unit, its operating income can only be expensed to pay overheads and personnel salaries, and any of surplus will be consolidated in a special accounts which is supervised by municipal council, thus the airport can be confined to be a public infrastructure. In China, after the confusion of chicken-and-egg priority problem years ago, the consensus to move faster to build GA airport network so as to promote GA development is reached, and more than 700 GA airports are planned to be newly constructed by 2020 in different areas. However, the facts are currently the formalities to obtain the approval of airports are still complicated; China has still not introduced the subsidy policies in terms of the construction and operation of GA airports; and the nature of public purpose may be challenged when the principal investment sources are privately-owned enterprises, therefore systematically, a system that guarantees public benefits shall be instituted.

## General Vocabulary

exotic　*adj.* 异国的；外来的；异国情调的
cluster　*n.* 群；簇；丛；串
urbanized　*adj.* 城市化的
theme　*n.* 主题；主旋律；题目
manufacture　*vt.* 制造；加工；捏造
residence　*n.* 住宅，住处；居住
recreation　*n.* 娱乐；消遣；休养
exhibition　*n.* 展览，显示；展览会；展览品
underpin　*vt.* 巩固；支持；从下面支撑；加强……的基础
full-fledged　*adj.* 全面发展的；经过全面训练的；成熟的；有充分资格的；羽毛生齐的
lifestyle　*n.* 生活方式
compatible　*adj.* 兼容的；能共处的；可并立的
geographical　*adj.* 地理的；地理学的
distinctive　*adj.* 独特的，有特色的；与众不同的

feature　　*n*. 特色，特征；容貌；特写或专题节目
generalize　　*vt*. 概括；推广；使……一般化
oriented　　*adj*. 以……为方向的；重视……的
headquarter　　*n*. 总部；指挥部；司令部
affection　　*n*. 喜爱，感情；影响；感染
vacation　　*n*. 假期；（房屋）搬出
livable　　*adj*. 适于居住的；生活过得有价值的；宜居的；值得一过的；可勉强在一起生活的；足够维持生活的；能对付，可处理（与liveable同）
amusement　　*n*. 消遣，娱乐；乐趣
versatile　　*adj*. 多才多艺的；通用的，万能的；多面手的
navy　　*n*. 海军，深蓝色的
airfield　　*n*. 飞机场
villa　　*n*. 别墅；郊区住宅
furnish　　*vt*. 提供；供应；装备
hangar　　*n*. 飞机库；飞机棚
fancy　　*adj*. 复杂的；昂贵的；精致的，花哨的；想象的；（食物）优质的；（花）杂色的；（动物）供观赏的
isolate　　*v*.（使）隔离，孤立；将……剔出；（某物质、细胞等）分离；区别看待（观点、问题等）
enclosure　　*n*. 附件；围墙；围场
upper-class　　*adj*. 上流社会的；上层阶级的；中学三年级、四年级的
wealthy　　*adj*. 富有的；充分的；丰裕的
celebrity　　*n*. 名人；名声
globally　　*adv*. 全球地；全局地；世界上
originally　　*adv*. 最初，起初；本来
emergence　　*n*. 出现，浮现；发生；露头
adequate　　*adj*. 充足的；适当的；胜任的
multitude　　*n*. 大量，多数；群众，人群
ex-service　　*adj*. 退役的
abandon　　*v*. 遗弃；离开；放弃；终止；陷入
surround　　*vt*. 围绕；包围
enormous　　*adj*. 庞大的，巨大的；凶暴的，极恶的
unprecedented　　*adj*. 空前的；史无前例的
correlated　　*adj*. 有相互关系的
dye　　*v*. 染；把……染上颜色；被染色
distribute　　*vt*. 分配；散布；分开；把……分类
attraction　　*n*. 吸引，吸引力；引力；吸引人的事物
mature　　*adj*. 成熟的；充分考虑的；到期的；成年人的
population　　*n*. 人口；种群，群体；全体居民；总体
expanse　　*n*. 宽阔；广阔的区域；苍天；膨胀扩张

vast　*adj.* 广阔的；巨大的；大量的；巨额的
florida　*n.* 佛罗里达（美国东南部的州）
municipal　*adj.* 市政的，市的；地方自治的
non-profit　*adj.* 非盈利的；不以盈利为目的的
offset　*v.* 抵消，弥补；衬托出；使偏离直线方向；用平版印刷术印刷，转印下一页；装支管
overheads　*n.* 企业的日常管理费用；杂项开支；一般费用（overhead 的复数）
personnel　*n.* 人事部门；全体人员
payroll　*n.* 工资单；在册职工人数；工资名单；工资
surplus　*n.* 剩余；顺差；盈余；过剩
deposit　*n.* 存款；押金；订金；保证金；沉淀物
supervision　*n.* 监督，管理
council　*n.* 委员会；会议；理事会；地方议会；顾问班子
strictly　*adv.* 严格地；完全地；确实地
startup　*n.* 启动；开办
hence　*adv.* 因此；今后
claim　*v.* 宣称；要求，索取；引起（注意）；获得；夺去（生命）；索赔（钱财）；需要
comprehensive　*adj.* 综合的；广泛的；有理解力的
real estate　不动产，房地产
extension　*n.* 延长；延期；扩大；伸展；电话分机
devoid　*adj.* 缺乏的；全无的
precondition　*n.* 前提；先决条件
ancillary　*adj.* 辅助的；副的；从属的
legal person　法人，法定代表人，法人代表
imperfectly　*adv.* 有缺点地；不完美地；未完成地
charge　*n.* 费用；电荷；掌管；控告；命令；负载
welfare　*n.* 福利；幸福；福利事业；安宁
FAA　*abbr.* 联邦航空管理局（Federal Aviation Administration）
subsidy　*n.* 补贴；津贴；补助金
facilitation　*n.* 简易化；助长；容易
aforementioned　*adj.* 上述的；前面提及的
consolidate　*vt.* 巩固，使固定；联合
confusion　*n.* 混淆，混乱；困惑
consensus　*n.* 一致；舆论；合意
formality　*n.* 礼节；拘谨；仪式；正式手续
complicated　*adj.* 难懂的，复杂的
principal　*adj.* 主要的；资本的
systematically　*adv.* 有系统地；有组织地
guarantee　*vt.* 保证；担保
institute　*v.* 实行，建立；授予……职位；提出（诉讼）

## Notes

1. Aviation town, an exotic concept, means a cluster of urbanized communities which is themed by the core business and infrastructure of general aviation, and holds a variety of functions such as manufacturing, residence, business, recreation, tourism and exhibition.

– a cluster of  一群；一组；一串
– a variety of  种种；各种各样的……

译文：航空小镇概念来自国外，就是围绕通航的核心业务和基础设施，具备生产、居住、商务、休闲、旅游会展等多种功能指向的城镇化聚集区。

2. And due to different geographical locations, and cultural and industrial background, aviation towns in foreign countries have been booming with distinctive features, which can be generalized into three types, i.e. industry-oriented, environment-oriented and activity-oriented.

– geographical location  地理位置，地理区位，地理环境
– be generalized into  归纳为
– oriented  以……为方向的，以……为取向的，以……为导向的

译文：国外航空小镇由于其不同的地理位置、文化产业背景，发展出各具特色的航空小镇，概况为三种类型：产业型、环境型和活动型。

3. Environment-oriented town draws people's affection by its environment and it can be grouped into tourism-and-vacation-specific one and livable-and-experiencing-specific one. Activity-oriented town gathers people by holding activities such as exhibitions, amusements and events.

– be grouped into  集合成，分成
– tourism-and-vacation-specific  旅游度假型的
– livable-and-experiencing-specific  宜居体验型的

译文：环境型以环境吸引人，包括旅游度假型和宜居体验型；活动型以活动聚集人，通过展会活动和娱乐赛事等聚集人气。

4. And the whole community is isolated from outside world by enclosure and also falls under 24-hour patrol, with primarily upper-class wealthy people and celebrities living inside. A 700m$^2$ villa, if connected directly to the runway and equipped with both hangar and swimming pool, could cost more than 2.5 million US dollars.

– isolate from  使隔离，使孤立
– upper-class  上流社会的；上层阶级的
– equipped with  装备着……；有……的配置
– swimming pool  游泳池

译文：整个社区设有围界与外界隔离并有24小时巡逻。社区内主要居住着处于社会上层的富豪及明星，一幢直接与跑道相连、机库泳池齐备、建筑面积700余平方米的别墅售价超过250万美元。

5. In the United States, there're also many fly-in communities at the service of aviation fans, and globally the rest of aviation towns are located in Europe, Australia, South America and so on.

— fly-in community　飞行社区

— at the service of　服务于

— South America　南美洲

译文：在美国，有大量主要服务飞行爱好者的飞行社区，其他分布在欧洲、澳大利亚、南美洲等。

6. After World War Ⅱ, multitudes of ex-service pilots gathered in many abandoned military airports, and over time, many fly-in communities surrounding such airports, or known as residence-oriented aviation town took shape.

— World War Ⅱ　第二次世界大战

— ex-service pilot　退役飞行员

— military airport　军用机场

— take shape　成形，形成，成型，有显著发展

译文：第二次世界大战后的美国大量废弃军用机场聚集了众多退役飞行员，久而久之衍变出很多以机场为中心的飞行爱好者社区——住宅型航空小镇。

7. Based on the enormous consuming market, the manufacturing industry of general aviation in the USA attained unprecedented development after World War Ⅱ, and gradually a town came into being, which was underpinned by the aviation manufacture and correlated industries.

— consuming market　消费市场

— unprecedented development　前所未有的发展

— come into being　形成，产生，存在，出现

— underpin by　巩固；支持；从下面支撑；加强……的基础

译文：第二次世界大战后通用航空制造业依托于其自身巨大的消费市场空前发展，逐渐形成以航空制造及关联产业为基础的小镇。

8. Among them, Florida enjoys the largest number of such aviation towns in the USA, i.e. about 70. As to GA airports in the United States, they have a wide range of property rights, including private, national, state, municipal and community ownership, and most of which are owned by governments at all levels though.

— the largest number of　数量最多的，数额最大的

— a wide range of　大范围的；许多各种不同的

— property right　产权

— at all levels　各个级别

译文：佛罗里达的航空小镇最多，约有70多个。美国通用机场的产权多种多样，既有私人拥有也有国家、州、市和社区拥有，多数通用机场归属各级政府所有。

9. As airport by nature is a non-profit unit, its operating income can only be allocated to offset overheads and personnel payrolls, and any of surplus will be deposited in a special accounts under the supervision of municipal council, which

then strictly confines the airport to be a public infrastructure.

- non-profit　非营利，非盈利，非营利性，非盈利性
- personnel payroll　人员薪资
- municipal council　市政委员会；市议会
- confine to　只限于；关在……里面

译文：机场本身是非盈利单位，其营业收入只能用于支付日常开销和人员工资，盈余部分则进入一个受市议会监管的特殊账户，这样机场被严格的限定为公共基础设施。

10. Nevertheless, there're nearly 200 regions claiming to have the aviation town in the pipeline, which lays a good foundation for the startup and advancement of aviation town in China. The aviation town extends the whole industrial chain of low-altitude tourism, forming a comprehensive tourism destination supported by a variety of businesses, including sightseeing, flight training, real estate development, extension of theme entertainments, and aviation festivals and events.

- claim to　要求，自称，索赔
- lays a good foundation　打下良好的基础
- low-altitude tourism　低空旅游
- real estate　房地产，不动产，基板面，房地产专业

译文：将近200个地区要做航空小镇，这为航空小镇在中国起步和发展奠定了一个良好基础。航空小镇将低空旅游的整个产业链条拉长，形成以游览观光、飞行培训、房地产开发、主题娱乐产品延伸、航空节事举办等各种产业为支撑的综合型旅游目的地。

11. Different from the aviation towns abroad, Chinese ones are devoid of specific developing orientation as well as clear development methods. Available airport, flight and airspace are the basic preconditions for the development of GA-featured town.

- different from　与……不同，不同于
- devoid of　没有，缺少，缺乏，全无的
- precondition　前提条件，先决条件

译文：与国外相比，中国式的航空小镇缺乏明确定位，发展途径的探索还不明晰。发展通用航空特色小镇，有机场、有飞行、有空域是基本的前提。

12. At present, only a small part of aviation towns have begun to take shape and the remaining are still in plan or under construction, for which the reasons are the weak legal-person-based GA manufacturing industry, restricted low-altitude airspace, imperfectly configured GA airports and flight-related infrastructure, as well as the lack of relevant supporting service and user basis for private flight.

- at present　目前，现在
- legal person　法人，法定代表人，法人代表

译文：目前，只有少部分航空小镇初具规模，大部分还处于规划和建设中。原因在于：我国法人通用航空制造业发展薄弱，低空空域管制制约，通用机场及飞行相关基础设施不健全，相关配套服务跟不上，缺乏私人飞行的群众基础。

13. And as far as the airport can satisfy the test of enough openness to the public and the low charges for public welfare consideration, such airport may apply to

FAA for AIP, a subsidy as high as 95% for the facilitation of its reconstruction and extension.

— as far as　至于，直到，远到；就……而言

— public welfare　公共福利，公用福利设施；社会福利

— FAA　联邦航空管理局（Federal Aviation Administration）是美国运输部下属负责民用航空管理的机构。联邦航空管理局和欧洲航空安全局（EASA, European Aviation Safety Agency）同为世界上主要的航空器适航证颁发者。联邦航空局的主要任务是保障民用航空的飞行安全，促进民航事业的发展，但不直接经营民航企业。联邦航空局的机构设置分总部、地区机构和地方机构三级。总部设在华盛顿，是国家的行政立法机构，负责制定民用航空的政策、规划和颁布规章制度、处理国际民用航空事务、领导本系统各地区和地方机构的工作。

译文：只要满足面向公众开放和公益性低收费的要求，机场就可以申请FAA的机场促进基金（AIP），为其改扩建提供高达95%的补贴支持。

# Exercises

Section one: Answer the following questions according to the text.
1. What is the definition of aviation town?
2. How many types does aviation town have?
3. What advantages does aviation tow enjoy?
4. Please give a brief introduction of fly-in community.
5. What historical reasons do affect the formation of aviation town in the US?

Section two: Match the words or phrases in column A with the definition in column B.

| A | B |
| --- | --- |
| 1. global | a. a structure at an airport where aircraft can be stored and maintained |
| 2. adequate | b. a prominent aspect of something |
| 3. runway | c. group of people willing to obey orders |
| 4. hangar | d. the basis on which something is grounded |
| 5. feature | e. involving the entire earth; not limited or provincial in scope |
| 6. headquarter | f. meeting the requirements especially of a task |
| 7. manufacture | g. a strip of level paved surface where planes can take off and land |
| 8. personnel | h. relating or belonging to or characteristic of a municipality |
| 9. municipal | i. the organized action of making of goods and services for sale |
| 10. foundation | j. the office that serves as the administrative center of an enterprise |

### Section three: Translate the following into Chinese.

1. Underpinned by the full-fledged GA industry and its rich GA culture, aviation town creates a lifestyle that is compatible with the development of general aviation.

2. Today in the Spruce Creek Airport, there're a 1,214-meter-long runway, two FBOs, more than 600 villas furnished with hangars and connected to airport runway, one championship-graded golf course and several fancy restaurants and clubs.

3. Originally from the United States, the aviation town has seen its emergence and further development in the country, thanks to the country's adequate airport resources and large number of pilots.

4. Compared with the residence-oriented aviation town, such industry-oriented town is dyed with strong industrial characteristics. As to eco-tour town, it is mainly distributed around the famous tourism attractions all over the world.

5. The mature GA industry and its mass population have made flying a lifestyle, and the vast expanse of lands, together with the specific living habits, shapes the form of towns and promotes their development.

6. With the progression of general aviation and the startup of private flight, the idea of aviation town has become more and more aware in China, and as a matter of fact, many cities plan to build or are building aviation towns.

7. And in China, most of aviation towns are planned and constructed surrounding the established GA airports, and some are constructed as aviation industry complex buildings which are based on the original aviation manufacturing industries and facilitated with relevant ancillary infrastructure.

### Section four: Translate the following into English.

1. 你过于信赖她的想法和专业知识了。
2. 她研究法国历史有很深的造诣。
3. 你的信用卡什么时候到期?
4. 申请表在格式和篇幅上差别很大。
5. 飞行途中,我们在芝加哥逗留了4个小时。
6. 这个城镇没多少娱乐活动——也就是这家影院和几家酒馆。
7. 这个演出季的演出节目包括5部新剧和几场中国和印度音乐的音乐会。
8. 政府已经限制出入境自由了。
9. 或许可以通过提高税费来限制汽车保有量的增长。
10. 他有3个需要照料的孩子。

# 2.2 Aviation culture
## 飞行文化

## Background

### Walmart and Its Aviation Culture

Airplanes enjoy one of the biggest benefits— the saving of time. Air traveling is unique because it is an air-based form of transportation, allowing passengers and cargo to travel in a more direct path between two points. It is common that some companies have used corporate aircraft to gain an advantage over their competitors in certain industries.

Walmart is one company that has used aviation from the beginning of the business to develop its business. As the founder of Walmart, Sam Walton once said: "Without business aircraft, the company would never have been successful."

It is well-known that Walmart is the largest retailer in the world, whose annual revenue reached about 482.1 billion USD in 2016. If Walmart were a country, it would rank 28th in world GDP, and have a larger economy than South Africa, UAE, and many other countries.

As the global largest private enterprise, Walmart has about 2.2 million employees in 11,000 stores over 27 countries around the world. Only the US Department of Defense and the Chinese Army have more employees than Walmart.

### Here is a brief history of the company

1945　At 26, Sam Walton bought his first variety store
1953　Sam Walton bought his first airplane, a used Ercoupe 4150C for $1,850
1969　Walmart started an Aviation Division to help grow its business
1970　38 stores, $44 million in sales, became a publicly traded company
1979　276 stores, $1 billion in sales
1982-1988　Forbes Magazine rank Sam Walton as the richest person in the US
1992　1,960 stores, 380,000 employees, and $43 billion in sales
Sam Walton passed away at the age of 74
1996　Walmart opened its first store in China in the city of Shenzhen
426 stores spreading out over 168 cities in China, almost $500 billion in sales worldwide

### Sam Walton's Passion for Flight

Barnard, one of Sam Walton's colleagues, shared this story: Walton was a very good pilot. He once pointed out one of his stores from the air with his face reflecting terror. He said: "There aren't enough cars in the parking lot!" And then, Walton put the plane into a nosedive. It was then I realized this was the end of my life and I said the man was absolutely insane. He was crazy. A couple of minutes later, we landed on what you would probably consider no more than a driveway; he got out of the plane and I followed him. Walton went inside the store to find a manager. The manager said the store stood empty because of a school festival that day, promising Walton the parking lot would fill later that afternoon. A friend once asked Sam Walton why he didn't just have one of his pilots fly for him, so he could take it easy. He replied to his friend: " If I let someone else, I'm going to lose that touch. I love to fly. And I don't want to lose that touch." As a pilot, it is essential to keep yourself in a higher physical and mental standard, following the rules of operation and maintaining good judgement. Operating a successful business also requires those skills.

## Aviation Division of Walmart

Since Sam Walton bought the first corporate airplane in 1953, Walmart has grown its own fleet which is one of the largest corporate aircraft fleet in the world. Sam Walton loved flying and he understood how to use airplanes as part of his company's business strategy.

Since there are no regional offices for Walmart, all 15,000 employees work in the headquarters in Bentonville, Arkansas. Beaver Lake Aviation, a Walmart subsidiary, operates the company's aviation business at the Rogers Municipal Airport, close to Walmart headquarters.

Here is a typical day in Beaver Lake Aviation at Rogers Airport. By 6:15 a.m., there are already about 100 passengers at Rogers Airport waiting to depart on a variety of jet aircraft. Most Learjets depart with 8 passengers (full load on a Walmart's Learjet) and fly throughout the country, making multiple stops by dropping off and picking up passengers.

The majority of passengers are front line managers, such as regional managers, buyers, human resource professionals and IT technicians, instead of the top management of Walmart. They visit the stores throughout the US and the world to assist the headquarters manage the business.

The average flight time of the fleets is 20,000 hours per year. There are 74 pilots flying an average of 55 hours each per month. In the matter of flight load factor, a typical corporate flight carries an average of 2.2 people per flight, or has a load factor of 2.2. Walmart corporate fleet has a load factor of 5.2, which means each Walmart flight carries an average of 5.2 passengers.

## Principle of Walmart's Aviation Culture–Don't overspend

Most big companies purchase brand-new airplanes as a symbol of their success. But all new aircraft, like most new machines, depreciate in value after purchased. It is secure to purchase and operate secondhand airplanes in good condition with a complete set of logbooks or records detailing every task done on the airplanes. These used airplanes can still maintain a sustaining value after a few years. Therefore, major corporate airplanes that Walmart has purchased have been used ones, not only because of their aviation culture is not to overspend, but also because the company always concentrates on low prices for their own products. Hence it is acceptable that Walmart doesn't spend a lot of money on brand-new fancy aircraft.

In contrast to most American large companies with many smaller offices all over the country, Walmart has only one office in the US and uses their aviation department to transport key personnel throughout the country. As airplanes have advantages of saving time and allowing people to communicate quickly, it is understandable that the corporate aircraft fleet of Walmart helps Walmart grow and operated its business more effectively and efficiently.

## General Vocabulary

retailer　*n.* 零售商；零售店
unique　*adj.* 独特的，稀罕的；唯一的，独一无二的
corporate　*adj.* 法人的；共同的，全体的；社团的；公司的；企业的
revenue　*n.* 收入；税收；销售额
spread out　扩展；铺开
division　*n.* 部门
competitor　*n.* 竞争者，对手
sales　*n.* 销售额；销售
publicly　*adv.* 公开地；公然地
pass away　逝世；停止
corporate　*adj.* 公司的；法人的；共同的
passion　*n.* 酷爱；热情；激情
strategy　*n.* 策略；战略
regional　*adj.* 地区的；分布的
founder　*n.* 创立者，创办者，创建者；浇铸工，制造金属铸件的人
headquarters　*n.* 总部；指挥部
depart　*vi.* 启程；离开
multiple　*a.* 许多的；多种多样的
well-known　众所周知的，出名的，知名的
drop off　送下车；减少
pick up　（开车等）接人；捡起
publicly　*adv.* 公然地；以公众名义
majority　*n.* 大多数
management　*n.* 管理人员；管理层
human resource　人力资源
colleague　*n.* 同事，同僚

professional   n. 专业人员；行家；专家
overspend   v. 超支；过度花费
brand-new   崭新的；全新的
symbol   n. 标志；象征；符号
depreciate   vi. 贬值；降价
festival   n. 节日；庆祝，纪念活动；欢乐
a set of   一套
maintain   vt. 维持；维修
sustaining   adj. 始终如一的；持续的；一致的
focus on   集中于；集中在
fancy   a. 昂贵的；时髦的
personnel   n. 全体职员；员工；人事部门
reflect   vt. 反射；反映
terror   n. 惊恐；惊慌失措
parking lot   停车场
absolutely   adv. 绝对地；完全地
insane   a. 疯狂的；精神错乱的；荒唐的
land on   降落于
effectively   adv. 有效地，生效地；有力地；实际上
driveway   n. 车道
take it easy   放轻松；别紧张
touch   n. 感觉；触摸
subsidiary   adj. 辅助的，次要的；附属的；子公司的；n. 子公司；辅助者
professional   n. 专业人员；职业运动员
assist   v. 参加，出席；协助（做一部分工作）；（通过提供金钱或信息）帮助；在场（当助手）；使便利；（在做某任务中）有助益
load factor   负荷系数，载荷因素
physically   adv. 身体上；物理上
mentally   adv. 精神上；心理上
brand   n. 品牌，商标；类型；烙印；（独特的）个性；烙铁；污名；燃烧的木头；（诗、文中的）剑
symbol   n. 象征；符号；标志
secondhand   adj. 二手的；旧的；间接获得的；做旧货生意的；adv. 间接地；间接听来；以旧货
detail   n. 细节，琐事；具体信息；次要部分；分队，支队；vt. 详述；选派
concentrate   vi. 集中；浓缩；全神贯注；聚集
operation   n. 运转；运行；操作
judgement（美 judgment）   n. 判断力；裁判
height   n. 高度；高处
acceptable   adj. 可接受的；合意的；可忍受的

efficiently　*adv*. 有效地；效率高地
Ercoupe 4150C　Ercoupe 4150C 型飞机（Ercoupe是最安全的单引擎飞机之一）
jet aircraft　喷气式飞机
full load　满载；满负荷
load factor　负载系数；（客机的）座位利用率
logbook　*n*. 飞行日志；航海日志
nosedive　*n*.（飞机）俯冲
cruising altitude　（飞机）巡航高度
South Africa　南非
UAE（United Arab Emirates）　阿拉伯联合酋长国
Department of Defense　国防部
Forbes Magazine　《福布斯》杂志
Bentonville　*n*. 本顿维尔（美国城市，沃尔玛总部所在地）
Arkansas　*n*.（美国）阿肯色州
Beaver Lake Aviation　比弗湖航空
Rogers Municipal Airport　罗杰斯市机场（阿肯色州）
Barnard　巴纳德

## Notes

1. Airplanes enjoy one of the biggest benefits- the saving of time. Air traveling is unique because it is an air-based form of transportation, allowing passengers and cargo to travel in a more direct path between two points.

- the saving of time　节省时间，time consuming　耗费时间的；旷日持久的
- air-based　基于航空的

译文：飞机最大的好处之一，就是节省时间。乘飞机出行的独特之处就是这一种基于飞机的运输方式，使乘客和货物在两点之间用最直接快捷的方式出行。

2. It is well-known that Walmart is the largest retailer in the world, whose annual revenue reached about 482.1 billion USD in 2016. If Walmart were a country, it would rank 28th in world GDP, and have a larger economy than South Africa, UAE, and many other countries.

- well-known　著名的；众所周知的；清楚明白的
- annual revenue　岁入；年度收入
- USD　*abbr*. 美元（USA dollar）
- GDP　*abbr*. 国内生产总值（Gross Domestic Product），国内生产总值（GDP）是指按国家市场价格计算的一个国家（或地区）所有常住单位在一定时期内生产活动的最终成果，常被公认为是衡量国家经济状况的最佳指标。国内生产总值GDP是核算体系中一个重要的综合性统计指标，也是我国新国民经济核算体系中的核心指标，它反映了一国（或地区）的经济实力和市场规模。
- UAE　*abbr*. 阿拉伯联合酋长国（United Arab Emirates），简称"阿联酋"，位于阿拉

伯半岛东部，北濒波斯湾，西北与卡塔尔为邻，西和南与沙特阿拉伯交界，东和东北与阿曼毗连，海岸线长734公里，总面积83600平方公里，首都阿布扎比。由阿布扎比、迪拜、沙迦、富查伊拉、乌姆盖万和阿治曼6个酋长国组成联邦国家。1972年2月10日，哈伊马角加入联邦。

译文：众所周知，沃尔玛是全世界最大的零售商，2016年的年收入达到4820亿美元。如果沃尔玛是一个国家，那它的国民生产总值可以排全世界第28位，并且是一个比南非、阿拉伯联合酋长国以及许多其他国家还要大的经济体。

3. As the global largest private enterprise, Walmart has about 2.2 million employees in 11,000 stores over 27 countries around the world. Only the US Department of Defense and the Chinese Army have more employees than Walmart. Forbes Magazine ranked Sam Walton as the richest person in the US from 1982 to 1988.

- private enterprise　民营企业；私营企业
- the US department of Defense　美国国防部
- the Chinese Army　中国军队
- Forbes Magazine　《福布斯》杂志，《福布斯》（Forbes）是美国一本商业杂志。该杂志每两周发行一次，以金融、工业、投资和营销等主题的原创文章著称。福布斯还报道技术、通信、科学和法律等领域的内容。福布斯杂志总部设于纽约市，它在美国商业类杂志的主要竞争对手是《财富》和《彭博商业周刊》。该杂志因其提供的列表和排名而为人熟知，包括最富有美国人列表（福布斯400）和世界顶级公司排名（福布斯全球2000）。福布斯杂志的座右铭是"资本家工具"（The Capitalist Tool），总编辑是史蒂夫·福布斯，行政总裁是麦克·佩里斯。

译文：作为全球最大的私企，沃尔玛在全球超过27个国家，11000家门店拥有大约220万名员工。只有美国国防部和中国军队拥有比沃尔玛更多的员工。从1982年到1988年，福布斯排行榜把山姆沃尔顿排为全美最富有的人。

4. As a pilot, it is essential to keep yourself in a higher physical and mental standard, following the rules of operation and maintaining good judgement. Operating a successful business also requires those skills.

- essential to　必不可少，必要的，对……非常重要的
- mental standard　心理状态
- maintain good judgement　保持良好的判断力

译文：作为一个飞行员，把身体和心理状态保持在一个高水准状态，遵循操纵守则，保持良好的判断力是很有必要的。经营一个成功的企业同样需要这些技能。

5. Since there are no regional offices for Walmart, all 15,000 employees work in the headquarters in Bentonville, Arkansas. Beaver Lake Aviation, a Walmart subsidiary, operates the company's aviation business at the Rogers Municipal Airport, close to Walmart headquarters.

- regional office　区域性公司，地区办事处，办事处，地区办公室
- Bentonville　本顿维尔（美国城市，沃尔玛总部所在地）
- close to　靠近，接近，接近于，在附近

译文：由于沃尔玛没有地区办事处，所有的15000名员工都工作在阿肯色州本顿维尔的总部。比弗湖航空是沃尔玛的一个子公司，在罗杰市立机场运营公司的航空业务，邻近沃尔玛

总部。

6. Here is a typical day in Beaver Lake Aviation at Rogers Airport. By 6:15 a.m., there are already about 100 passengers at Rogers Airport waiting to depart on a variety of jet aircraft. Most Learjets depart with 8 passengers (full load on a Walmart's Learjet) and fly throughout the country, making multiple stops by dropping off and picking up passengers.

— a variety of　种种；各种各样的……
— Learjet　里尔喷射机，是美国盖茨·利尔喷气公司研制的双发轻型行政机
— drop off　减少；让……下车；睡着
— pick up　捡起；获得；收拾；（汽车；飞机）乘载；不费力地学会

译文：这是罗杰机场比弗湖航空的典型一天。早晨6点15分之前，已经有大约100名乘客在罗杰机场等待各种各样喷气式飞机起飞。大多数里尔喷气机搭乘8名乘客，飞行穿越美国，中途有多个站点放下和接乘客。

7. The majority of passengers are front line managers, such as regional managers, buyers, human resource professionals and IT technicians, instead of the top management of Walmart.

— majority of　大多数；大部分
— human resource　人力资源，人事部
— top management　高管理层；董事会

译文：大多数乘客是前线经理，比如地区经理、买手、人力资源专家和IT技术人员，而不是沃尔玛的高层管理人员。

8. In the matter of flight load factor, a typical corporate flight carries an average of 2.2 people per flight, or has a load factor of 2.2. Walmart corporate fleet has a load factor of 5.2, which means each Walmart flight carries an average of 5.2 passengers.

— corporate flight　公务飞行
— load factor　负载系数，满载率，载荷因子。负载系数，是指航空器承运的旅客数量与航空器可提供的座位数之比，它反映了航空器座位的利用程度，是体现航班效益和空运企业经济效益的重要指标。

译文：关于飞行的负载系数，一个典型的公务机平均每趟搭乘2.2名乘客，或者说负载系数是2.2。沃尔玛公务机机队的负载系数是5.2，意味着每次沃尔玛的飞行平均搭乘5.2名乘客。

9. Most big companies purchase brand-new airplanes as a symbol of their success. But all new aircraft, like most new machines, depreciate in value after purchased. It is secure to purchase and operate secondhand airplanes in good condition with a complete set of logbooks or records detailing every task done on the airplanes.

— secondhand airplane　二手飞机
— in good condition　处于良好的状态，状态良好，身体健康
— a set of　一套；一组；一副

译文：大多数大公司会买全新的飞机来彰显他们的成功。但是全新的飞机，就像大多数全新的机器一样，购买过后就会贬值。购买和运营条件良好，有全套的飞行日志或详尽记录每次飞行任务的二手飞机是很安全的。

10. In contrast to most American large companies with many smaller offices all over the country, Walmart has only one office in the US and uses their aviation department to transport key personnel throughout the country.
- In contrast to　相比之下，成对比
- key personnel　主要工作人员；关键员工

译文：相较于大多数美国其他大公司在全美有很多分公司，沃尔玛在美国只有一个公司，利用他们的航空部门在美国境内运送关键的管理人员。

## Exercises

Section one: Answer the following questions according to the text.
1. Why did Sam Walton found Walmart's aviation department?
2. Is Sam Walton a great pilot? Give some reasons to prove your opinions.
3. Please describe a typical day in Beaver Lake Aviation at Rogers Airport.
4. What is the principle of Walmart's aviation culture?
5. Unlike other big companies in the US, why does Walmart buy used airplanes instead of brand-new ones?

Section two: Match the words or phrases in column A with the definition in column B.

| A | B |
| --- | --- |
| 1. overspend | a. a merchant who sells goods at retail |
| 2. logbook | b. not plain; decorative or ornamented |
| 3. secondhand | c. an associate you work with |
| 4. retailer | d. the legal document stating the reasons for a judicial decision |
| 5. management | e. spend more than available of (a budget) |
| 6. depreciate | f. a book in which the log is written |
| 7. fancy | g. previously used or owned by another |
| 8. colleague | h. strong feeling or emotion |
| 9. judgement | i. the act of managing something |
| 10. passion | j. lose in value |

Section three: Translate the following into Chinese.
1. Airplanes enjoy one of the biggest benefits- the saving of time. Air traveling is unique because it is an air-based form of transportation, allowing passengers and cargo to travel in a more direct path between two points.
2. It is well-known that Walmart is the largest retailer in the world, whose annual revenue reached about 482.1 billion USD in 2016. If Walmart were a country, it would rank 28th in world GDP, and have a larger economy than South Africa, UAE, and many

other countries.

3. As a pilot, it is essential to keep yourself in a higher physical and mental standard, following the rules of operation and maintaining good judgement. Operating a successful business also requires those skills.

4. Since Sam Walton bought the first corporate airplane in 1953, Walmart has grown its own fleet which is one of the largest corporate aircraft fleet in the world. Sam Walton loved flying and he understood how to use airplanes as part of his company's business strategy.

5. The average flight Time of the fleets is 20,000 hours per year. There are 74 pilots flying an average of 55 hours each per month.

6. These used airplanes can still maintain a sustaining value after a few years. Therefore, major corporate airplanes that Walmart has purchased have been used ones, not only because of their aviation culture is not to overspend, but also because the company always concentrates on low prices for their own products.

### Section four: Translate the following into English.
1. 人很容易对安眠药产生依赖性。
2. 我能不能上大学要看我能考多少分。
3. 市政委员会鼓励开发该房地产项目，旨在促进就业和增加娱乐设施。
4. 不论我想做什么，他们总是会鼓励我。
5. 本届文化节将涵盖从音乐、戏剧、芭蕾到文学、电影、视觉艺术等各种艺术形式。
6. 我的任何想法都无法付诸实践，这让我很灰心。
7. 自己在家做饭的人似乎越来越少了。
8. 他们拒绝透露是如何得到我的地址的。
9. 她因病无法出席为她新近推出的小说举办的庆祝聚会。
10. 因天气情况不佳，宇宙飞船的发射被推迟了24个小时。

# 2.3
# GA airport in USA
# 美国通用航空机场建设

## Background

In the United States, all airports can be defined as general aviation airport given that they are popularly engaged by general aviation, and particularly those hub airports in the metropolises are proposed to be named as transport airport. There're 19,100 airports across the country, out of which some 17,800 airports are registered in FAA, 651 airports are operated by public air transportation, and the remaining nearly 18,400 airports are operated either by general aviation or by individuals. Small airports in the United States are efficiently utilized because the master plan of airports is dedicated to making for the diversion of air traffic at peak hours. When the demand for GA flight arises, airport information published by FAA may be referred to consult the situation and capability of any airports in service. Generally airports owned by government and organizations are more stable and sustainable, and in particular those airports constructed under the funding of federal governments are reassuringly solid.

### Air traffic control service in USA

Today's fast and easy GA flight in the United States should give credits to its soundly developed air navigation system, which has gone through eight-decade development since the system was established. In the 1920s, pilots used to rely on the radio station and rotational light to fly from one land to the other, but due to the limited coverage of the signal light, pilot could be lost while navigating in the zone. Ten years later, the federal government turned to introduce navigation aids to provide services for pilots day and night. And the radio communication in both the calm and stormy weather can attain coverage at four bands, i.e. one set of equipment was capable of transmitting radio signal to four different directions. Funded by the government, the network equipment was allocated to provide navigation for pilots during their whole journey.

In company with the development of general aviation, VHF Ominidirectional Range (VOR) also comes into being. Today, based on above 1,000 VORs installed one after another in the airports for GA service, these VORs have been developed into an increasingly sounder coverage network where 360 central call radar stations and central navigation stations throughout the United Stations jointly transmit a huge pattern of round wheel to direct pilots to land on a runway in airport, even if under bad weather. In addition, Instrument Landing System (ILS) is the most widely used equipment which enables pilots to achieve approaching and safe landing under various climatic conditions and FAA has deployed more than 1,100 ILSs in airports across the United States. For GA flight, air traffic control system is in a crucial position since it oversees the order of aircrafts so that they can avoid each other and orderly fly in and out of the airport terminal areas. In America, traffic control centers have complete equipment and functions in which each of equipment undertakes varied task and the controllers in the control tower and terminal area handle respectively the process of take-off and landing of flights. There're about 700 airports set up with control towers and relevant controllers all around the USA, and except hub airports, most of which are to serve general aviations. The rushed control tower in airport is equipped with control apparatuses respectively in charge of safety and monitoring of air traffic flow. Particularly about 300 airports have established surveillance radars for general aviation in order to provide assistance to GA pilots when they are in need of so.

Another useful service is to provide a traffic control network, commonly known as 24-hour air route traffic control center (ARTCC) for general aviation flight. These centers provide radars to monitor the air traffic flow and render flight planning services within the corresponding air zones fallen under their control. Nevertheless, the flight activities of general aviation are complicated and the guidance of instruments has limits, therefore it is pilot himself who shall call the shot when the radar becomes out of reach during the visible flight or cross-clouds descending of aircraft. For the air transport tasks, especially regular commuter flights undertaken by general aviation, it is always under the monitoring of instrument, be the weather good or bad.

### Development policy of GA in USA

The development research and policy support for GA in the United States are briefly undergone as follows:

(1) In 1926, the Air Commerce Act was issued, and the early regulations on general aviation were formulated.

(2) On August 12, 1994, President Clinton signed the General Aviation Revitalization Act, and with that, the United States vigorously tapped the transportation function of general aviation, making the private aircrafts the third pillar carrier of air transportation just after the large aircrafts for major air routes and the regional aircrafts.

(3) Since 1994, an array of plans and proposals for the development and research of GA was rolled out one after another, including General Aviation Propulsion (GAP), Aviation Safety and Aviation Meteorological Information Proposal, Highway in the Sky (HITS) and Small Aircraft Transportation System (SATS). It laid out a number of goals: within 10 years, small aircrafts could reach 25% of the country's urban, rural and remote destinations at a quadruple speed over automobiles and within 25 years, they could reach 90% of communities; general aviation, a more or less door-to-door air transport should become a constituent part of individualized transport system in which airworthy light aircrafts could fly among more than 10,000 nationwide small airports opened nearly under all weather and be an important alternative of airline aircraft and automobiles in people's travels, so as to relieve the traffic on interstate highways and hub airports. Americans named the above developmental goals as "National General Aviation Blueprint" in which the SATS and AGATE (Advanced General Aviation Transport Experiments) were in the most significant roles.

(4) The industry's self-regulations have been soundly developed. A good number of associations or organizations have been self-established by the members of GA

industry, which in chronicle order includes National Association of State Aviation Officials, American Association of Airport Executive, Aircraft Owners and Pilot Association, Helicopter Association International, National Agricultural Aviation Association, National Aviation Training Association, Aviation Distributor and Manufacturers Association, National Business Aircraft Association, National Association of Flight Trainer, Small Aircraft Manufacturers Association, United States Ultralight Association, the General Aviation Manufacturers Association and regional airline associations, etc. These organizations actively engage in technical exchanges, industry management and self-regulation within the industrial scope, and request member enterprises to carry out business activities in accordance with various regulations of the country. Meanwhile, the associations also, on behalf of the interests of each member, communicate with and drum up support from the government, share reasonable advises, propagandize general aviation and study the development plan of the industry, so as to attract the attention of the government and the society to general aviation and promote the healthy development of general aviation.

(5) In 1992, the plan of Young Eagles was launched. The goal of the plan was to introduce one million young people to the world of flight by the centennial turn of aviation industry in December, 2003. The plan popularized aviation industry among mass population and nurtured and inspired people's passion for flight and the exploration for nature and sky. At the same time, the plan also enhanced America's grasp on aviation flight and training of relevant talents, making the country more competitive in global stage and guaranteeing its leading position in international aviation flight. In 2002 alone, the plan set a record high by attracting 115,000 people to flight circle and made people engaging in flight at a total accumulative number of 875,000. In 2003, the plan was successfully delivered.

## General Vocabulary

particularly  *adv.* 异乎寻常地；特别是；明确地
hub  *n.* 中心；毂；木片
metropolis  *n.* 大都市；首府；重要中心
propose  *vt.* 建议；打算，计划；求婚
efficiently  *adv.* 有效地；效率高地（efficient 的副词形式）
utilize  *vt.* 利用
known  *adj.* 知道的；闻名的；已知的
dedicate  *vt.* 致力；献身；题献
diversion  *n.* 转移；消遣；分散注意力
peak  *n.* 山峰；最高点；顶点；帽舌
arise  *vi.* 出现；上升；起立
consult  *vt.* 查阅；商量；向……请教

generally　adv. 通常；普遍地，一般地
federal　adj. 联邦的；同盟的；联邦政府的；联邦制的
reassuringly　adv. 安慰地；鼓励地
solid　adj. 固体的；可靠的；立体的；结实的；一致的
credit　n. 信用，信誉；贷款；学分；信任；声望
rely　vi. 依靠；信赖
rotational　adj. 转动的；回转的；轮流的
signal　n. 信号；暗号；导火线
zone　n. 地带；地区；联防
introduce　vt. 介绍；引进；提出；采用
aid　n. 援助；帮助；助手；帮助者
stormy　adj. 暴风雨的；猛烈的；暴躁的
transmit　vt. 传输；传播；发射；传达；遗传
direction　n. 方向；指导；趋势；用法说明
network　n. 网络；广播网；网状物
journey　n. 旅行；行程
instal　vt. 安装；使就任；设置
increasingly　adv. 越来越多地；渐增地
jointly　adv. 共同地；连带地
enable　v. 使能够；使成为可能；授予权利或方法；（计算机）启动
climatic　adj. 气候的；气候上的；由气候引起的；受气候影响的
deploy　vt. 配置；展开；使疏开
crucial　adj. 重要的；决定性的；定局的；决断的
oversee　vt. 监督；审查；俯瞰；偷看到，无意中看到
orderly　adv. 顺序地；依次地
rushed　adj. 匆忙的；贸然的；v. 急忙（rush的过去式）；匆促
apparatus　n. 装置，设备；仪器；器官
monitor　n. 监视器；监听器；监控器；显示屏；班长
radar　n. 雷达，无线电探测器
assistance　n. 援助，帮助；辅助设备
route　n. 路线，航线；道路，公路；（交通工具的）固定路线；巡访；途径，渠道；（北美）递送路线；用于美国干线公路号码前
corresponding　adj. 相当的，相应的；一致的；通信的
guidance　n. 指导，引导；领导
instruments　n. 仪器（instrument的复数）；工具；乐器
visible　adj. 明显的；看得见的；现有的；可得到的
descending　v. 下降；下倾；降临；下（坡，楼梯）；依次递降；（声响）渐低；（感觉，气氛）突然笼罩；交由……继承（descend 的现在分词）
commuter　n. 通勤者，经常乘公共车辆往返者；月季票乘客
undergo　vt. 经历，经受；忍受

issue　　*vt.* 发行，发布；发给；放出，排出
vigorously　　*adv.* 精神旺盛地，活泼地
tap　　*v.* 轻敲；装上嘴子；窃听；采用；在树上切口；委任；非法劝说转会
pillar　　*n.* 柱子，柱形物；栋梁；墩
array　　*n.* 数组，阵列；排列，列阵；大批，一系列；衣服
propulsion　　*n.* 推进；推进力
highway　　*n.* 公路，大路；捷径
lay out　　展示；安排；花钱；为……划样；提议
quadruple　　*adj.* 四倍的；四重的
automobile　　*n.* 汽车
door-to-door　　*adj.* 挨家挨户的；送货上门的
constituent　　*adj.* 构成的；选举的；有任命（或选举）权的；立宪的；有宪法制定（或修改）权的
individualized　　*adj.* 个人的；有个性的；具有个人特色的
airworthy　　*adj.* 适宜航空的；耐飞的；飞机性能良好的
nationwide　　*adj.* 全国范围的；全国性的
alternative　　*n.* 二中择一；供替代的选择
relieve　　*vt.* 解除，减轻；使不单调乏味；换……的班；解围；使放心
interstate　　*adj.* 州际的；州与州之间的
blueprint　　*n.* 蓝图；行动方案；（生物细胞的）模型
soundly　　*adv.* 酣畅地；明智地；牢固地；不错地；严厉地；完全地
association　　*n.* 协会，联盟，社团；联合；联想
chronicle　　*n.* 编年史，年代记；记录
National Association of State Aviation Officials 国家航空官员协会
American Association of Airport Executive 美国机场行政人员协会
Aircraft Owners and Pilot Association 飞机拥有者和飞行员协会
Helicopter Association International 国际直升机协会
National Agricultural Aviation Association 国家农业航空协会
National Aviation Training Association 国家航空培训协会
Aviation Distributor and Manufacturers Association 航空经销商和制造商协会
National Business Aircraft Association 国家公务机协会
National Association of Flight Trainer 国家飞行教官协会
Small Aircraft Manufacturers Association 小飞机制造商协会
United States Ultralight Association 美国超轻型飞机协会
the General Aviation Manufacturers Association and regional airline associations 通用航空制造商协会和地区航空公司协会
exchange　　*n.* 交换；交流；交易所；兑换
self-regulation　　*n.* 自我调节，自律，自我调控
request　　*vt.* 要求，请求
accordance　　*n.* 按照，依据；一致，和谐

通用航空专业英语

drum up　招徕（顾客）；竭力争取；纠集；鼓动
propagandize　*vt.* 宣传；对……进行宣传
centennial　*adj.* 一百年的
popularize　*vt.* 普及；使通俗化
nurture　*vt.* 养育；鼓励；培植
grasp　*n.* 抓，握；理解，领会；力所能及，把握；权力，控制
talent　*n.* 才能；天才；天资
deliver　*vt.* 交付；发表；递送；释放；给予（打击）；给……接生

## Notes

1. In the United States, all airports can be defined as general aviation airport given that they are popularly engaged by general aviation, and particularly those hub airports in the metropolises are proposed to be named as transport airport.

- hub airport　枢纽机场，中心航空港
- propose to　建议，向某人求婚，打算

译文：按照美国的定义，所有的机场均可称为通用航空机场，因为它们普遍被通用航空使用，某些大都市的枢纽机场则被提议称作运输机场。

2. Small airports in the United States are efficiently utilized because the master plan of airports is dedicated to making for the diversion of air traffic at peak hours. When the demand for GA flight arises, airport information published by FAA may be referred to consult the situation and capability of any airports in service.

- master plan　总体规划，总平面，主计划
- dedicated to　奉献给，献身于，致力于，专门为
- peak hours　高峰时间

译文：美国小机场的使用率高，机场总体规划便于飞行高峰的分流。通用航空飞行需要时可从FAA发布的机场情报上查询到任何正在使用的机场的状况和保障能力。

3. In the 1920s, pilots used to rely on the radio station and rotational light to fly from one land to the other, but due to the limited coverage of the signal light, pilot could be lost while navigating in the zone. Ten years later, the federal government turned to introduce navigation aids to provide services for pilots day and night.

- rely on　依靠，依赖
- radio station　无线电台，广播电台
- federal government　联邦政府
- day and night　日日夜夜，夜以继日，没日没夜

译文：20世纪20年代，当飞行员由无线电发射站和旋转信号灯光引导，从一个着陆地飞行到另一个着陆地时，由于旋转信号灯的覆盖范围有限，飞行员在飞行区域的导航有断档区。10年后，联邦政府开始引入导航辅助设备，日夜为飞行员提供服务。

4. In company with the development of general aviation, VHF Ominidirectional Range (VOR) also comes into being. Today, based on above 1,000 VORs installed one

after another in the airports for GA service, these VORs have been developed into an increasingly sounder coverage network where 360 central call radar stations and central navigation stations throughout the United Stations jointly transmit a huge pattern of round wheel to direct pilots to land on a runway in airport, even if under bad weather.

— VHF *abbr.* 甚高频（Very High Frequency），甚高频通信系统应用于航空事业中，是保证在飞行的过程中飞机与地面之间以及飞机与飞机之间能够保持相互联系的通信工具。

— VOR VHF omnidirectional radio range，甚高频全向信标。是指一种工作于112～118MHz，可在360°范围内给航空器提供它相对于地面台磁方位的近程无线电导航系统，一种用于航空的无线电导航系统。其工作频段为112～118兆赫的甚高频段，故此得名。VOR通常与测距仪（Distance Measuring Equipment, DME）同址安装，在提供给飞行器方向信息的同时，还能提供飞行器到导航台的距离信息，这样飞行器的位置就可以唯一的被确定下来。VOR在北美大陆和欧洲大陆分布广泛，可以覆盖整个欧美大陆。通常，这些信标的部署与主要航线的航路点和测距仪站点的交叉点分布一致，这样可以利用航线网络/信标结构对飞机的全程飞行进行导航。

译文：随着通用航空的发展，甚高频全向信标仪（VOR）应运而生。有1000余个VOR陆续安装在机场为通用航空服务。今天，VOR已经在原来的基础上形成了愈发完善的覆盖网络，全美360个话台中心雷达站和导航中心站向外发射一个巨大的圆轮图案，引导飞行员在特定的航路上准确地飞行。导航设备还帮助飞行员从航路上下降高度，引导飞行员在机场某条跑道着陆，即使在天气状况不好的情况下也能做到这一点。

5. In addition, Instrument Landing System (ILS) is the most widely used equipment which enables pilots to achieve approaching and safe landing under various climatic conditions and FAA has deployed more than 1,100 ILSs in airports across the United States.

— in addition 除此之外，另外

— Instrument Landing System (ILS) 仪器降落系统，盲降系统，是应用最为广泛的飞机精密进近和着陆引导系统。它的作用是由地面发射的两束无线电信号实现航向道和下滑道指引，建立一条由跑道指向空中的虚拟路径，飞机通过机载接收设备，确定自身与该路径的相对位置，使飞机沿正确方向飞向跑道并且平稳下降高度，最终实现安全着陆。

— approaching 进近，进近是指飞机下降时对准跑道飞行的过程，在进近阶段，要使飞机调整高度，对准跑道，从而避开地面障碍物，飞行员必须要把注意力高度集中才能准确操作，因此进近是有着严格的标准和操作规程的。现代商业航空运输主要是以大型客机为主来进行的。大型客机主要体现在飞机吨位大、速度大、安全责任大。因此有一种安全可行的辅助着陆系统来减轻飞行员的操纵负荷，提高飞行的安全性是必须的。而且由于天气的能见度的问题，这种系统可以说对航班运输的经济性安全性也是至关重要的，因此产生了仪表着陆系统。

译文：仪表着陆系统（ILS）是应用最广泛的设备，它使飞行员实现各种气候条件下的进近飞行和安全着陆。FAA在全美机场安装了1100多个ILS系统。

6. There're about 700 airports set up with control towers and relevant controllers

all around the USA, and except hub airports, most of which are to serve general aviations. The rushed control tower in airport is equipped with control apparatuses respectively in charge of safety and monitoring of air traffic flow.

— set up  建立；装配；开业；竖立

— control tower  （机场）指挥塔台，控制塔

— in charge of  主管，负责，在掌管之下，看管

译文：全美有近700个机场有塔台控制，除了枢纽机场外，塔台控制机场主要为通用航空所使用。繁忙的机场塔台安装有负责安全、观察空中交通流的控制仪器。

7. Another useful service is to provide a traffic control network, commonly known as 24-hour air route traffic control center (ARTCC) for general aviation flight. These centers provide radars to monitor the air traffic flow and render flight planning services within the corresponding air zones fallen under their control.

— air route traffic control center（ARTCC）  航路交通控制中心

— traffic flow  交通流量

— flight planning  飞行计划

译文：另一项有用的服务是为通用航空飞行提供交通控制网络，一般称为24h空中航路交通控制中心（ARTCC）。这些中心提供的雷达负责监控空中飞机流量，分别在控制的空中区域提供飞行计划服务。

8. Since 1994, an array of plans and proposals for the development and research of GA was rolled out one after another, including General Aviation Propulsion (GAP), Aviation Safety and Aviation Meteorological Information Proposal, Highway in the Sky (HITS) and Small Aircraft Transportation System (SATS).

— an array of  一排；一批；大量

— roll out  铺开，滚出，推出（新产品、服务等），实行（新制度），开展

译文：1994年以后，相继推出了"通用航空推进"（GAP）计划、"航空安全和航空气象信息"计划、"空中高速公路"（HITS）计划和"小飞机运输系统"（SATS）计划等一系列通用航空发展研究计划。

9. General aviation, a more or less door-to-door air transport should become a constituent part of individualized transport system in which airworthy light aircrafts could fly among more than 10,000 nationwide small airports opened nearly under all weather and be an important alternative of airline aircraft and automobiles in people's travels, so as to relieve the traffic on interstate highways and hub airports.

— more or less  或多或少，差不多

— door-to-door  门到门，挨家挨户地，送货上门

— interstate highway  州级公路

译文：使通用航空这种近乎门到门的空中运输成为个人化运输系统的一个组成部分，让易于飞行的轻型飞机在近乎全天候开放的全国10000多个小型机场间飞行，成为人们出行中部分代替航线飞机和汽车的一种重要选择，从而缓解拥挤的州际高速公路和枢纽机场的压力。

10. A good number of associations or organizations have been self-established by the members of GA industry, which in chronicle order includes National Association of

State Aviation Officials, American Association of Airport Executive, Aircraft Owners and Pilot Association, Helicopter Association International, National Agricultural Aviation Association, National Aviation Training Association, Aviation Distributor and Manufacturers Association, National Business Aircraft Association, National Association of Flight Trainer, Small Aircraft Manufacturers Association, United States Ultralight Association, The General Aviation Manufacturers Association and regional airline associations, etc.

译文：通用航空工业的各成员都有其自行组成的协会或组织，先后成立的协会有：国家航空官员协会、美国机场管理协会、飞机拥有者和飞行员协会、国际直升机协会、国家农业航空协会、国家航空培训协会、航空经销商和制造商协会、国家公务机协会、国家飞行教官协会、小飞机制造商协会、美国超轻型飞机协会、通用航空制造商协会和地区航空公司协会等。

11. Meanwhile, the associations also, on behalf of the interests of each member, communicate with and drum up support from the government, share reasonable advises, propagandize general aviation and study the development plan of the industry, so as to attract the attention of the government and the society to general aviation and promote the healthy development of general aviation.

— on behalf of　代表，为了

— drum up　招徕（顾客）；竭力争取；纠集；鼓动

— so as to　以便，为了，为的是，使得

译文：同时，协会代表各方会员的利益与政府进行沟通和游说，提出合理建议，宣传通用航空，研究行业发展规划，以此引起政府和社会对通用航空的关注，促进通用航空健康发展。

12. At the same time, the plan also enhanced America's grasp on aviation flight and training of relevant talents, making the country more competitive in global stage and guaranteeing its leading position in international aviation flight.

译文：同时"雏鹰计划"也增强了美国在航空飞行方面的实力和人力储备资源，提高了国家竞争力，保证了其在国际上的航空飞行领先地位。

## Exercises

Section one: Answer the following questions according to the text.

1. What is the commonly recognised definition of general aviation airport?
2. What services does air traffic control in USA provide ?
3. What functions do VOR and ILS have?
4. As another useful service, what advantages and disadvantages does ARTCC enjoy?
5. Please briefly describe the history of development policy of general aviation policy in USA.

Section two: Match the words or phrases in column A with the definition in column B.

A  B
1. issue                  a. something intended as a guide for making something else
2. formulate              b. move downward and lower, but not necessarily all the way
3. research               c. a device that requires skill for proper use
4. blueprint              d. a formal organization of people or groups of people
5. automobile             e. transmit information
6. association            f. a motor vehicle with four wheels
7. communicate            g. systematic investigation to establish facts
8. instrument             h. prepare and issue for public distribution or sale
9. radar                  i. a pulse of microwave radiation is used to detect distant objects
10. descend               j. elaborate, as of theories and hypotheses

Section three: Translate the following into Chinese.

1. There're 19,100 airports across the country, out of which some 17,800 airports are registered in FAA, 651 airports are operated by public air transportation, and the remaining nearly 18,400 airports are operated either by general aviation or by individuals.

2. Generally airports owned by government and organizations are more stable and sustainable, and in particular those airports constructed under the funding of federal governments are reassuringly solid.

3. Today's fast and easy GA flight in the United States should give credits to its soundly developed air navigation system, which has gone through eight-decade development since the system was established.

4. Funded by the government, the network equipment was allocated to provide navigation for pilots during their whole journey.

5. For GA flight, air traffic control system is in a crucial position since it oversees the order of aircrafts so that they can avoid each other and orderly fly in and out of the airport terminal areas.

6. For the air transport tasks, especially regular commuter flights undertaken by general aviation, it is always under the monitoring of instrument, be the weather good or bad.

7. Within 10 years, small aircrafts can reach 25% of the country's urban, rural and remote destinations at a quadruple speed over automobiles and within 25 years, they can reach 90% of communities.

8. In 1992, the plan of Young Eagles was launched. The goal of the plan was to introduce one million young people to the world of flight by the centennial turn of

aviation industry in December, 2003.

Section four: Translate the following into English.
1. 这家航空公司下个月将开辟一条新的跨越大西洋的航线。
2. 我们快走到大厅时可以听到里面传出的笑声。
3. 广告公司总是得想出新办法来促销。
4. 绿色和平组织致力于促使人们了解当前威胁地球的危机。
5. 她刚刚被提升为高级销售代表。
6. 这间餐厅棒极了，不过从外面看可一点都看不出来。
7. 当地居民对停车位短缺一事非常恼火。
8. 这家旅馆的酒吧只对其住客开放。
9. 她是这所大学意大利文学方面的常驻专家。
10. 在过去的两年里，我们的销售额一直在逐步上升。

# 2.4 Status of GA in USA
# 美国通用航空现状

## Background

### Development history of GA in the world

General aviation is one of the critical components of civil aviation. It is born and developed along with the emergence and development of civil aviation. On December 17, 1903, the aircraft invented by Wright Brothers in the United States was successfully flown, which ushered in a new era of modern aviation and also raised curtain on the development of general aviation in the world.

In the first half of the 20th century, the world underwent great changes, and especially the outbreak of the two world wars had a profound and positive impact on the development of aviation technology around the world. During World War Ⅰ when airplanes were put to use in the war, some countries became aware of the military significance of aircrafts and successively set up aviation science and technology research institutions, thus the aviation industry system was set in motion. From the 1920s, significant changes took place in the performance and structure of aircrafts. Namely, the aircrafts were restructured from double wing to cantilever single wing, from wooden structure to fully metallic structure, from open cockpit to closed cockpit and from fixed landing gear to retractable landing gear, and the power of the aircraft engine were also increased fivefold. Thanks to these scientific and technological achievements, the flight speed of aircrafts was quickened by 2 up to 4 times, and the aviation industry gradually grew into an independent industrial sector. The earliest transport aircraft was the DC series manufactured by USA-based Douglas Aircraft Manufacturing Company whose DC-2 aircraft, tested in World War I, enjoyed a proven technology and could basically answered the requirements for safety, reliability, comfort and economy. And at the time, the power of the piston-engined DC-2 aircraft was considerably enhanced in a bid to respond to the call for a faster flight speed.

### Status of GA in USA

After World War II ended, aviation technology was rapidly developed and a great many military aircrafts were transferred into civil use. As a result, general aviation gained momentum of growth and finally in 1970s flourished. However, at the beginning of 1980s, as the global economy was set back and the GA aircraft market was saturated while the technical innovations decreased, GA began its downturn and eventually hit rock bottom. Since the 1990s, in company with the continuous growth of the world economy and the introduction of encouraging policies by governments as well as a range of new aviation products, general aviation revived and rose again with great potentials to tap.

According to the statistics in 2004, there were about 340,000 general-purpose aircrafts in service all around the world, of which the United States accounted for about two thirds, i.e. about 211,000 aircrafts, making the North America the most developed region of general aviation in the world. Also, almost 90% of all general-purpose aircrafts worldwide were made in the United States. And the massive fleets of GA aircrafts were primarily distributed in the United States, Canada, Europe and other developed countries. In addition, Brazil, a developing country in South America, also had a relatively well developed GA industry as it possessed more than 20,000 general-purpose aircrafts, and in Russia, their general-purpose aircrafts were totaled at a number of above 10,000.

**The status quo of general aviation in the United States will be highlighted as follows.**

(1) The role of GA in USA's economic growth

Today, general aviation in the United States has been so developed that it, together with the transport aviation, comprises the country's safest and most efficient air transportation system. It holds together the overall balance of civil aviation transportation industry, and bridges United States with the rest of the world.

GA also lays the foundation for air transport. In America, aviation services are convenient and practical for almost everyone, because air transport carriers and GA companies play in full separately independent but collectively harmonious role in the transportation service. GA's presence is mainly outside the metropolises, and with its flexible and specialized methods, it transports annually millions of passengers as well as tens of thousands of tons of cargo and it delivers mails and packages to more remote regions at the shortest time. Meanwhile, flights of GA are also operated every now and then in the airports of certain major cities in the United States. It is known that during the economic construction of the country, GA, together with transport aviation, contributes a total of 20% of its GDP.

Speaking of the main reasons for the rapid development of general aviation, the Americans would attribute the achievements made in the past 2 centuries to their close interplay with the changes and innovations of all sorts in air transport. Firstly, credits should be taken by America's deregulation policy in 1970s. Secondly the industrial development gives impetus to the growth of GA. Many factories and supply centers are initially based in or near the metropolises before the GA could become the new powerhouse for national transport, but then these manufacturing industries were collectively relocated from the densely populated cities to the outlying townlets in which many GA airports were built and GA started its burgeoning.

(2) Business line and scale of GA in USA

By now, though still under the revival stage, GA industry in the United States remains a large industrial sector and involves actively in all lines of trade and people's lives. And by and large, GA in the country can be grouped into six types of aviation activities as follows. First and foremost, business flight takes up a large proportion of general aviation activities, and it constitutes a huge and extensive service system of business aviation; then the aircraft leasing and short-haul flight also take a good part in the GA activities. People with pilot license rent aircraft from the airport service provider (FBO) to provide charter services to the public or to carry his own families, friends and colleges, which is just as convenient as car rent. In America, there're now about 2,000 general-purpose aircrafts owned by regular short-haul airline companies as well as another 6,000 aircrafts owned by aircraft leasers and charter flight operators, and it is well established that short-haul airlines have become an indispensable air link between small cities and major air hub cities in the United States; as to agricultural, forestry and animal husbandry flight, it is an important means of developing American agriculture industry on a yearly basis, approximately more than two billion acres of arable lands are farmed with GA; then, the private flight is also one of the major business lines of general aviation in the United States. There're 600,000 Americans who hold private flight licenses and are also used to flying small airplanes to engage in business or recreational flight activities; additionally, the aviation sports makes up 5% of all GA activities across the United States, and as a matter of fact, a good number of universities and colleges have their own flight sports teams; last but not least, the aviation training is deemed as an enormous line of business in America, whose growth status is a barometer of the country's GA, and virtually one out of four pilots in the

America's airlines is trained through such GA training programs.

As at the end of 2004, the GA aircrafts in the United States were totaled at a number of 220,000 or a proportion of 71.7% of the total such aircrafts worldwide. Moreover, a total of 28.12 million hours of GA flights was flown and there were around 619,000 people holding valid pilot licenses, including more than 385000 general aviation pilots. In the year of 2004, GA industry directly contributed USD 41 billion to the U.S. economy, indirectly created USD 102 billion worth of values, and provided nearly 600,000 jobs for the society.

## General Vocabulary

component　　*n.* 组成部分；成分；组件，元件
Wright Brothers　　莱特兄弟（飞机发明者）
era　　*n.* 时代；年代；纪元
raise　　*vt.* 提高；筹集；养育；升起；饲养，种植
curtain　　*n.* 幕；窗帘
outbreak　　*n.*（战争的）爆发；（疾病的）发作
profound　　*adj.* 深厚的；意义深远的；渊博的
positive　　*adj.* 积极的；正的，阳性的；确定的，肯定的；实际的，真实的；绝对的
impact　　*n.* 影响；效果；碰撞；冲击力
technology　　*n.* 技术；工艺；术语
World War Ⅰ　　第一次世界大战
military　　*adj.* 军事的；军人的；适于战争的
successively　　*adv.* 相继地；接连着地
institution　　*n.* 制度；建立；（社会或宗教等）公共机构；习俗
motion　　*n.* 动作；移动；手势；请求；意向；议案
structure　　*n.* 结构；构造；建筑物
wing　　*n.* 翼；翅膀；飞翔；派别；侧厅，耳房，厢房
cantilever　　*n.* 悬臂
wooden　　*adj.* 木制的；僵硬的，呆板的
metallic　　*adj.* 金属的，含金属的
cockpit　　*n.* 驾驶员座舱；战场
fixed　　*adj.* 确定的；固定的；处境……的；准备好的
gear　　*n.* 齿轮；装置，工具；传动装置
landing gear　　起落架；起落装置，着陆装置
retractable　　*adj.* 可缩进的；可收起的；伸缩自如的
engine　　*n.* 引擎，发动机；机车，火车头；工具
fivefold　　*adv.* 五倍地；五重地
quicken　　*vi.* 加快；变活跃；进入胎动期
independent　　*adj.* 独立的；单独的；无党派的；不受约束的

basically　adv. 主要地，基本上
reliability　n. 可靠性
comfort　n. 安慰；舒适；安慰者
piston　n. 活塞
considerably　adv. 相当地；非常地
respond　vi. 回答；作出反应；承担责任
transfer　v. 转让；转接；移交；转移（地方）；（使）换乘；转存，转录；调动（工作）；传染，传播；使（运动员）转队；把（钱）转到另一账户，机构上
set back　推迟；使……受挫折；把……往回拨
saturate　vt. 浸透，使湿透；使饱和，使充满
innovation　n. 创新，革新；新方法
downturn　n. 衰退（经济方面）；低迷时期
eventually　adv. 最后，终于
bottom　n. 底部；末端；臀部；尽头
continuous　adj. 连续的，持续的；继续的；连绵不断的
revive　vi. 复兴；复活；苏醒；恢复精神
tap　v. 轻敲；装上嘴子；窃听；采用；在树上切口；委任；非法劝说转会
massive　adj. 大量的；巨大的，厚重的；魁伟的
primarily　adv. 首先；主要地，根本上
developed country　发达国家
possess　vt. 控制；使掌握；持有；迷住；拥有，具备
total　vt. 总数达
comprise　vt. 包含；由……组成
bridge　vt. 架桥；渡过
practical　adj. 实际的；实用性的
separately　adv. 分别地；分离地；个别地
collectively　adv. 集体地，共同地
presence　n. 存在；出席；参加；风度；仪态
package　n. 包，包裹；套装软件，程序包
attribute　vt. 归属；把……归于
interplay　n. 相互影响，相互作用
innovation　n. 创新，革新；新方法
deregulation　n. 放松管制，解除管制，放宽管制，撤销管制
impetus　n. 动力；促进；冲力；动量
powerhouse　n. 精力充沛的人，身强力壮的人；强大的集团（或组织）；强国；权势集团；权威人士；动力源
relocate　vt. 重新安置；迁移
densely　adv. 浓密地；密集地
populate　vt. 居住于；构成人口；移民于；殖民于
outlying　adj. 边远的；无关的

townlet　　*n.* 小镇；小城镇
revival　　*n.* 复兴；复活；苏醒；恢复精神；再生效
actively　　*adv.* 积极地；活跃地
proportion　　*n.* 比例，占比；部分；面积；均衡
constitute　　*vt.* 组成，构成；建立；任命
lease　　*n.* 租约；租期；租赁物；租赁权
short-haul　　*adj.*（尤指空运）短途运输的
indispensable　　*adj.* 不可缺少的；绝对必要的；责无旁贷的
acre　　*n.* 土地，地产；英亩
arable　　*adj.* 适于耕种的；可开垦的
additionally　　*adv.* 此外；又，加之
deem　　*vt.* 认为，视作；相信
barometer　　*n.* 气压计；晴雨表；显示变化的事物
virtually　　*adv.* 事实上，几乎；实质上

## Notes

1. On December 17, 1903, the aircraft invented by Wright Brothers in the United States was successfully flown, which ushered in a new era of modern aviation and also raised curtain on the development of general aviation in the world.

－ Wright Brothers　莱特兄弟是美国著名的发明家，哥哥是威尔伯·莱特（Wilbur Wright，1867年4月16日—1912年5月12日），弟弟是奥维尔·莱特（Orville Wright，1871年8月19日—1948年1月30日）。1903年12月17日，莱特兄弟首次试飞了完全受控、依靠自身动力、机身比空气重、持续滞空不落地的飞机，也就是世界上第一架飞机"飞行者一号"。

译文：1903年12月17日，美国莱特兄弟发明的飞机飞行成功，开创了现代航空的新纪元，同时也揭开了世界通用航空发展的序幕。

2. During World War I when airplanes were put to use in the war, some countries became aware of the military significance of aircrafts and successively set up aviation science and technology research institutions, thus the aviation industry system was set in motion.

－ put to use　使用；利用
－ aware of　意识到，知道
－ set in motion　开始，调动；把……发动起来

译文：第一次世界大战期间，由于飞机在战争中的应用，一些国家政府开始注意到了飞机的军事意义，相继成立了航空科学技术研究机构，航空工业体系初见端倪。

3. From the 1920s, significant changes took place in the performance and structure of aircrafts. Namely, the aircrafts were restructured from double wing to cantilever single wing, from wooden structure to fully metallic structure, from open cockpit to closed cockpit and from fixed landing gear to retractable landing gear, and the power

of the aircraft engine were also increased fivefold.

— take place　发生；举行；进行；产生

译文：从20世纪20年代开始，飞机的性能和构造发生了巨大的变化：由双机翼飞机发展到张臂式单机翼飞机；由木质结构飞机发展到全金属结构飞机；由敞开式座舱飞机发展到密闭式座舱飞机；由固定式起落架飞机发展到收放式起落架飞机。

4. The earliest transport aircraft was the DC series manufactured by USA-based Douglas Aircraft Manufacturing Company whose DC-2 aircraft, tested in World War Ⅰ, enjoyed a proven technology and could basically answered the requirements for safety, reliability, comfort and economy.

— Douglas Aircraft Manufacturing Company　道格拉斯飞机制造公司，麦克唐纳-道格拉斯公司（McDonnell-Douglas Corporation）是美国制造飞机和导弹的大垄断企业。1939年由詹姆斯·麦克唐纳创办，称麦克唐纳飞机公司。1967年兼并道格拉斯飞机公司，改为现名。从那时起，与波音公司在1997合并后，它生产了一些知名的商业和军用飞机，如DC-10客机和F-15鹰等空中优势战斗机。该公司总部设在密苏里州的圣路易斯附近的路易斯国际机场，而其子公司麦道公司的总部则设在未合并的密苏里州圣路易斯县。

— World War Ⅰ　第一次世界大战，是在19世纪末至20世纪初，资本主义国家向其终极阶段，即帝国主义过渡时产生的广泛的不可调和矛盾，战争过程主要是同盟国和协约国之间的战斗。德意志帝国、奥匈帝国、奥斯曼帝国、保加利亚王国属于同盟国阵营，大英帝国、法兰西第三共和国、俄罗斯帝国、美利坚合众国和意大利王国等国则属于协约国阵营。这场战争是欧洲历史上破坏性最强的战争之一。第一次世界大战给人类带来了深重灾难，但在客观上促进了科学技术的发展。在第一次世界大战中，各种新式武器如飞机、毒气、坦克、远程大炮相继投入战争，是武器发展史的重要阶段。

译文：最早的运输飞机是美国道格拉斯飞机制造公司生产的DC产品系列。经过第一次世界大战的洗礼，美国的DC-2飞机在技术上已经比较成熟，基本上达到了安全、可靠、舒适和经济的要求。

5. However, at the beginning of 1980s, as the global economy was set back and the GA aircraft market was saturated while the technical innovations decreased, GA began its downturn and eventually hit rock bottom.

— set back　推迟；使……受挫折；把……往回拨

— technical innovation　科技创新，技术革新，技术创新，技术改造

— hit rock bottom　陷入低谷，感到绝望（俚语）

译文：进入80年代后，由于全球性的经济衰退、通用航空飞机数量的相对饱和、技术创新减少，导致通用航空开始下滑并陷入低谷。

6. Since the 1990s, in company with the continuous growth of the world economy and the introduction of encouraging policies by governments as well as a range of new aviation products, general aviation revived and rose again with great potentials to tap.

— encouraging policy　鼓励性政策

— a range of　一系列；一些；一套

译文：90年代以来，随着世界经济的持续增长、各国政府出台鼓励政策、航空产品推陈出新，通用航空又呈现复苏和重新崛起的态势，且有强劲后势。

通用航空专业英语

7. Today, general aviation in the United States has been so developed that it, together with the transport aviation, comprises the country's safest and most efficient air transportation system. It holds together the overall balance of civil aviation transportation industry, and bridges United States with the rest of the world.

- civil aviation　民用航空

译文：美国通用航空发展到今天，已经与运输航空一起组成了美国最安全、最有效的航空运输系统，维系着民航运输业的总体平衡，编织着美国与世界的交通桥梁网络。

8. GA's presence is mainly outside the metropolises, and with its flexible and specialized methods, it transports annually millions of passengers as well as tens of thousands of tons of cargo and it delivers mails and packages to more remote regions at the shortest time.

- tens of thousands of tons of　成千上万吨
- at the shortest time　在最短的时间内
- delivers mails and packages　派送邮件和包裹

译文：通用航空主要活跃在大都市以外，以其机动灵活的特有方式每年运输百万旅客、千万吨货物，以最快的速度将邮件送达更偏远的地区。

9. Speaking of the main reasons for the rapid development of general aviation, the Americans would attribute the achievements made in the past 2 centuries to their close interplay with the changes and innovations of all sorts in air transport.

- speaking of　说到，谈及
- attribute to　归因于，把……归功于，归咎于

译文：谈起通用航空快速发展的主要原因，美国人总结说：事实上，美国200年间所取得的成就与各种不同形式的空中运输方式的变革、创新紧紧联系在一起。

10. Many factories and supply centers are initially based in or near the metropolises before the GA could become the new powerhouse for national transport, but then these manufacturing industries were collectively relocated from the densely populated cities to the outlying townlets in which many GA airports were built and GA started its burgeoning.

- powerhouse　n. 精力充沛的人，身强力壮的人；强大的集团（或组织）；强国；权势集团；权威人士；动力源
- densely populated　人口稠密的

译文：在通用航空成为国家运输生力军之前，许多工厂和供应中心都位于大都市或其附近的地方，后来，制造工业集中迁移出了人口稠密的城市，到边远的小城镇地区，因此，在那里修建了许多通用航空机场，通用航空也随之快速发展起来。

11. Then the aircraft leasing and short-haul flight also take a good part in the GA activities. People with pilot license rent aircraft from the airport service provider (FBO) to provide charter services to the public or to carry his own families, friends and colleges, which is just as convenient as car rent.

- short-haul flight　短途飞行，long-haul flight　长途飞行
- take a good part in　在……方面起主动作用

080

- FBO 以固定机场为基地的飞行活动（fixed-base operation），位于机场或者邻近机场的为通用航空飞机、公务机和私人飞机提供停场、检修、加油、清洁、休息等服务的基地或服务商。FBO 的服务对象也主要是通用航空飞机，特别是公务机和私人飞机。FBO 通航企业性质是综合性的通用航空服务企业或基地，提供停场、检修、加油等服务，业务范围比较广泛。除了飞机的维护、维修外，还包括飞机销售、租赁和飞行培训等综合服务，给私人飞机客户提供全方位立体的服务。

译文：租赁飞机和短航线飞行在通航活动中不可小视，持有飞行驾照的人从机场服务商（FBO）那里租用飞机，向公众提供包机服务，或运送自己的家人、朋友、同事，其方便程度就像租用汽车一样。

12. Last but not least, the aviation training is deemed as an enormous line of business in America, whose growth status is a barometer of the country's GA, and virtually one out of four pilots in the America's airlines is trained through such GA training programs.

译文：航空培训在美国是一个巨大的产业，其繁荣程度是美国通用航空的晴雨表，美国各航空公司每四名驾驶员中就有一人是经过通用航空培训的。

## Exercises

### Section one: Answer the following questions according to the text.

1. What is the business line and scale of GA in USA?
2. What is the role of GA in USA's economic growth?
3. Please briefly describe the development history of GA.
4. Except the US, what other countries have a relatively large industry of GA?
5. Please discuss the importance of Wright Brothers in the development histroy of GA.

### Section two: Match the words or phrases in column A with the definition in column B.

| A | B |
| --- | --- |
| 1. component | a. worthy of reliance or trust |
| 2. era | b. a period marked by distinctive character from a fixed point or event |
| 3. cockpit | c. a state of equilibrium |
| 4. reliable | d. be brought back to life, consciousness, or strength |
| 5. eventually | e. an abstract part of something |
| 6. possess | f. furnish with people |
| 7. balance | g. have ownership or possession of |
| 8. metropolis | h. compartment where the pilot sits while flying the aircraft |
| 9. populate | i. after a long period of time or an especially long delay |
| 10. revive | j. people living in a large densely populated municipality |

Section three: Translate the following into Chinese.

1. First and foremost, business flight takes up a large proportion of general aviation activities, and it constitutes a huge and extensive service system of business aviation

2. As to agricultural, forestry and animal husbandry flight, it is an important means of developing American agriculture industry on a yearly basis, approximately more than two billion acres of arable lands are farmed with GA

3. In the year of 2004, GA industry directly contributed USD 41 billion to the U.S. economy, indirectly created USD 102 billion worth of values, and provided nearly 600,000 jobs for the society.

4. Meanwhile, flights of GA are also operated every now and then in the airports of certain major cities in the United States. It is known that during the economic construction of the country, GA, together with transport aviation, contributes a total of 20% of its GDP.

5. In addition, Brazil, a developing country in South America, also had a relatively well developed GA industry as it possessed more than 20,000 general-purpose aircrafts, and in Russia, their general-purpose aircrafts were totaled at a number of above 10,000.

6. According to the statistics in 2004, there were about 340,000 general-purpose aircrafts in service all around the world, of which the United States accounted for about two thirds, i.e. about 211,000 aircrafts, making the North America the most developed region of general aviation in the world.

7. In the first half of the 20th century, the world underwent great changes, and especially the outbreak of the two world wars had a profound and positive impact on the development of aviation technology around the world.

Section four: Translate the following into English.
1. 当你一路往南走下去的时候，你就会注意到气候的细微变化。
2. 午饭后我们在会议室集合。
3. 道路施工导致市中心交通阻塞。
4. 他接受过昂贵的教育，但却继续做一名服务员。
5. 除工资以外，我还有退休金和医疗保险。
6. 图书馆的书是按照学科分类的。
7. 我喜欢他服装设计中简洁典雅的线条
8. 兼职工作可以作为从不工作过渡到全职工作的桥梁。
9. 一家德国公司与一家瑞士小公司合作开发该产品。
10. 英国警方和意大利警方通力合作，抓住了恐怖分子。

# Unit 3

# GA aircrafts and manufacture
# 通用航空飞机及制造

# 3.1 Classifications of GA aircraft
## 通用航空飞机机型分类

## Background

By international practice, general aviation refers to all aircrafts, flight scope and activities excluding air force and passenger and cargo transportation of civil aviation. It is a well-developed and sophisticated system that can provide various flight-related air services for individuals and institutions. Whether you want to charter an airplane to go sightseeing with families in a foreign place, or customize a business aircraft for your company, or buy a privately owned means of transportation in the air, a big toy that can carry and delight yourself, all of it can be quick-fixed easily and smoothly as long as you have the idea ready and it is related with flight. Following is the introductions of several common models of general aviation aircraft in China.

### 1. Robinson R-44

Robinson R-44 helicopter is one of the best sold light-duty helicopters in the global market, and is also the most used helicopter in China's general aviation. There're now 200 such helicopters in China which are mainly applied to pilot training program, private flight, commercial flight and so on.

Parameter data: the R-44 helicopter totals 11.76 meters long, and its fuselage is 2.1 meters wide and 3.28 meters high. The cruise speed is 216 km/h; the gross weight is 1,090 kg; the maximum climbing altitude is 4,270 meters; the maximum range (no reserve) is approximately 643 km; the cockpit can accommodate one pilot and three passengers, and the helicopter is powered by Lycoming IO-540 EFI engine. The standard price is 3.92 million yuan (including tax).

### 2. H125[AS350B3e] Squirrel, Airbus Helicopter

The H125 [AS350B3e] helicopter is a six-seat single-engine helicopter featuring solid and durable performance, high reliability and low cost. With more than 5,400 worldwide sales done, it is now the best sold light-duty single-engine helicopter in China. By its strong driving force, complete service range and excellent highland performance, it outranks other equivalents. And there're about 90 such helicopters in China, which are mainly used for emergency medical care, agrochemical spraying, electric power inspection, firefighting, etc.

Parameter data: the H125 helicopter is 12.94 meters long, 2.53 meters wide, 3.34

meters high, with cruise speed at 251 km/h, maximum weight at 2,800 kg, maximum climb altitude at 5,258 meters and maximum range at 686 km, and with 1+5/1+6 seats and Turbomeca's Arriel 2D turboshaft engine. The complete machine is sold at 20 million yuan.

### 3. Robinson R-22

The R-22 Beta Ⅱ helicopter, by the strength of its superior performance, proven safety, affordable price and the most cost-efficient operation, keeps every record in the performance of speed, altitude and flight range among helicopters of same class, and also is accepted as the most popular entry-level lightweight helicopter in the world. There're 70 such helicopters in China, whose main application is in private flight, beginner's flight and pilot license training.

Parameter data: R-22 helicopter has the total length at 6.58 meters, fuselage by 1.93 meters wide and 2.71 meters high, cruise speed at 190 km/h, maximum takeoff weight at 621 kilograms, maximum climbing altitude at 4,267 meters, maximum range at 480 kilometers, seats by 1+1, and a four-cylinder engine typed as the Lycoming O-360. The reference price is 3 million yuan.

### 4. 407GX (Bell 407GX)

Thanks to the advantageous combination of speed, performance and maneuverability, Bell407 helicopter has established an outstanding market acceptance. Its spacious passenger cabin can carry 7 passengers and can be flexibly configured within the payload range in accordance with the business requirements. Its multiple power-driven components and outperformed flight control can ensure a comfortable piloting and boarding experience in various environments. By record there're 50 such helicopters and they may execute general transportation, official transportation, offshore oil or gas exploitation support, medical rescue, search and rescue, air law enforcement, helicopter pilot training, liaison, aerial observation, armed reconnaissance and other missions.

Parameter data: Bell 407GX helicopter has the total length at 12.7 meters, fuselage by 2.68 meters wide and 3.56 meters high, cruise speed at 246 km/h, maximum takeoff weight at 2,722 kilograms, maximum climbing altitude at 5,157 meters and maximum range at 611 kilometers, together with 7 seats and a Rolls-Royce 250-C47B/8 engine. The reference price is 22 million yuan.

### 5. Agusta AW119KX

The AW119KX helicopter, an eight-seat single-engine helicopter under Agusta Westland, Italy, dominates out of question the single-engine helicopter industry by its maximum take off weight of 3 tons and its leading performance. With the flexible operation features, it can be applied widely in every field of general aviation, and be reconfigured quickly according to the requirements of missions. In comparison with other models of helicopter of the same class, it enjoys larger payload, longer range and cruise time as well as a cabin so spacious that makes it a high-end model which can load two pairs of stretchers and a single-engine helicopter truly capable of emergency medical service (EMS). It is low in operational cost, long in time interval of regular inspection, and easy in maintenance. It can satisfy various operational requirements of different customers, and it is mainly used for medical emergency rescue, forest protection and fire prevention, patrol and alarm, and satellite hot spot reconnaissance.

Parameter data: the Agusta AW119KX helicopter has the total length at 13.01 meters, rotating diameter at 10.83 meters, fuselage by 3.60 meters high, cruise speed at 257 km/h, maximum takeoff weight at 2,850 kg, maximum climb altitude at 4,572 meters and maximum range at 954 km, together with 8 seats and a Pratt-Whitney PT6B-37A engine. The reference price is 29.77 million yuan.

### 6. Enstrom 480B

Enstrom 480 helicopter, a lightweight multi-purpose five-seat civil helicopter developed by the USA-based Enstrom Helicopter Corporation, is updated according to the US Army's requirements on the new training helicopter, hence it is reputed for its safety, stability, comfort and good value. There're about 20 such helicopters in China, which are mainly used for coach training, official flight, cargo transportation, law enforcement and patrol, reconnaissance, etc.

Parameter data: the Enstrom 480B helicopter has the total length at 9.10 meters, diameter of rotor at 9.75 meters, fuselage by 3.0 meters high, cruise speed at 202 km/h, maximum takeoff weight at 1361 kg, practical service ceiling up to 3,962 meters and maximum range at 685 km, together with 5 seats and a Rolls-Royce 250-C20W engine. The reference price is 9 million yuan.

### 7. Schwarzer S-300C

Schwarzer S-300C is a world-renowned multi-purpose helicopter by its performance, reliability and cost-efficiency. It combines perfectly the maneuverability, load capacity, range and flight endurance, and it is safe, reliable, easy to pilot and of good stability. There're currently a total of about 3,000 S-300C helicopters in service across the world, and they serve in the field of agrochemical spraying, cargo transportation, personal flight, aerial photography, law enforcement and patrol, power and traffic inspection, etc.

Parameter data: the S-300C helicopter has the total length at 9.4 meters, wingspan at 8.2 meters, fuselage by 2.65 meters high, cruise speed at 159 km/h, maximum takeoff weight at 930 kg, practical service ceiling up to 4,450 meters and maximum range at 763 kilometers, together with 3 seats and a Lai Kangming HIO-360-D1A engine. The reference price is 3.2 million yuan.

### 8. Bell 206L4

Bell 206 helicopter is probably the first impression many people have on helicopters just as if Boeing 747 once stood for the very impression of passenger aircraft. It is supposedly accepted as a classic work in the industrial design and people in its field comment it as the "safest and most reliable" helicopter. Such comment means more its being an industry standard than a mere title of honor. With its widespread global presence, it has exerted far-reaching impact on the development of civil turboshaft helicopter. The Bell 206L4 helicopter, also known as the "Long Range Assault", is a seven-seat general-purpose helicopter developed on the basis of Bell 206B "Jet Assault" and improved from the classic 206B-3 model. It is a multi-mission helicopter with the best-combined power, space and range. And it has been, by now, the safest, most reliable and most mature light turboshaft helicopter, with the main application to missions such as business charter flights, emergency medical rescue, flight training, aerial photography, surveying and mapping, administrative services and law enforcement flights.

Parameter data: Bell 206L4 is 12.92 meters long, 2.38 meters wide, 3.32 meters high, with cruise speed at 206 km/h, maximum take-off weight at 2018 kg, operative hover altitude at 3,048 meters, maximum range at about 662 km, seats by 1+6 or 1+7 as well as a turboshaft engine typed as Rolls-Royce 250-C30P. The reference price is 20 million yuan.

## General Vocabulary

sophisticated   *adj.* 复杂的；精致的；久经世故的；富有经验的
delight   *vt.* 使高兴
smoothly   *adv.* 平稳地，平滑地；顺利地，流畅地；平静地；均匀地
light-duty   *adj.* 轻型的
parameter   *n.* 参数；系数；参量
fuselage   *n.* 机身（飞机）
cruise speed   巡航速度
gross weight   毛重，总重量
maximum   *adj.* 最高的；最多的；最大极限的
climbing   *n.* 爬山；攀登；上升

accommodate　vt. 容纳；使适应；供应；调解
standard price　标准价格
tax　n. 税金；重负
squirrel　n. 松鼠；松鼠毛皮
durable　adj. 耐用的，持久的
performance　n. 性能；绩效；表演；执行；表现
reliability　n. 可靠性
highland　adj. 高原的；高地的
outrank　vt. 地位高于，级别高于；居……之上位
equivalent　n. 对等的人（或事物）；当量
electric power　电力；电功率
firefighting　n. 消防；防火，打火
kilometer　n. 千米，公里
kilogram　n. 千克，公斤
range　n. 范围；幅度；排；山脉
turboshaft　n. 涡轮轴；涡轮轴发动机
engine　n. 引擎，发动机；机车，火车头；工具
superior　adj. 上级的；优秀的，出众的；高傲的
affordable　adj. 负担得起的
cost-efficient　adj. 有成本效益的；合算的
entry-level　adj. 入门的；初级的；最低阶层的；适合于初学者的
beginner　n. 初学者；新手；创始人
four-cylinder engine　四缸发动机
advantageous　adj. 有利的；有益的
combination　n. 结合；组合；联合；化合
maneuverability　n. 可操作性；机动性
outstanding　adj. 杰出的；显著的；未解决的；未偿付的
acceptance　n. 接纳；赞同；容忍
spacious　adj. 宽敞的，广阔的；无边无际的
cabin　n. 小屋；客舱；船舱
payload　n.（导弹、火箭等的）有效载荷，有效负荷；收费载重，酬载；（工厂、企业等）工资负担
in accordance with　依照；与……一致
outperform　vt. 胜过；做得比……好
execute　vt. 实行；执行；处死
enforcement　n. 执行，实施；强制
liaison　n. 联络；连音
reconnaissance　n. 侦察；勘测（与reconnoissance同）；搜索；事先考查
stretcher　n. 担架；延伸器
capable　adj. 有能力的；有才干的；容许……的；可以做（某事）的；综合性的；有资格的

operational cost　　经营成本；操作费用
alarm　　*n.* 闹钟；警报，警告器；惊慌
satellite　　*n.* 卫星；人造卫星；随从；卫星国家
rotate　　*vi.* 旋转；循环
diameter　　*n.* 直径
repute　　*vt.* 名誉；认为；把……称为
stability　　*n.* 稳定性；坚定，恒心
coach　　*n.* 教练；旅客车厢；长途公车；四轮大马车
rotor　　*n.* 转子；水平旋翼；旋转体
service ceiling　　飞行高度；实用升限；升高限度
renowned　　*adj.* 著名的；有声望的
load capacity　　负载能力，载重能力
endurance　　*n.* 忍耐力；耐久性；持续时间
power　　*n.* 力量，能力；电力，功率，性能；政权，势力；幂
wingspan　　*n.* 翼展；翼幅
impression　　*n.* 印象；效果，影响；压痕，印记；感想；曝光
supposedly　　*adv.* 可能；按照推测；恐怕
classic　　*adj.* 经典的；古典的，传统的；最优秀的
comment　　*vt.* 发表评论；发表意见
widespread　　*adj.* 普遍的，广泛的；分布广的
exert　　*vt.* 运用，发挥；施以影响
far-reaching　　*adj.* 深远的；广泛的；伸至远处的
turboshaft　　*n.* 涡轮轴；涡轮轴发动机
assault　　*n.* 攻击；袭击
surveying and mapping　　测绘科学与技术，测绘学
hover　　*vi.* 盘旋，翱翔；徘徊
medical　　*adj.* 医学的；药的；内科的

## Notes

1. Whether you want to charter an airplane to go sightseeing with families in foreign place, or customize a business aircraft for your company, or buy a privately owned means of transportation in the air, a big toy that can carry and delight yourself, all of it can be quick-fixed easily and smoothly as long as you have the idea ready and it is related with flight.

– as long as　　只要，如果

译文：无论你是要包一架飞机和家人去异地观光，还是为企业定制一架用于公务的飞机，或者买一架属于自己的空中交通工具，一个可以载着自己游戏的大玩具，只要有想法，只要和飞行相关，这一切的一切都可以简单而轻松、快捷和顺畅办到。

2. Robinson R-44 helicopter is one of the best sold light-duty helicopter in the

global market, and is also the most used helicopter in China's general aviation.

— light-duty  轻型的，轻的

— most used  最常用的，应用最广泛的

译文：罗宾逊R44 R-44直升机是全球轻型直升机市场上销量最大的直升机之一，同时也是国内通航最多的直升机。

3. By its strong driving force, complete service range and excellent highland performance, it outranks other equivalents. And there're about 90 such helicopters in China, which are mainly used for emergency medical care, agrochemical spraying, electric power inspection, firefighting, etc.

译文：该机以强劲的动力、全面的业务能力、卓越的高原性能，领先同级别直升机。国内有90架左右，主要用途有紧急医疗救护、喷洒农药、电力巡查、消防等。

4. The R-22 Beta Ⅱ helicopter, by the strength of its superior performance, proven safety, affordable price and the most cost-efficient operation, keeps every record in the performance of speed, altitude and flight range among helicopters of same class, and also be accepted as the most popular entry-level lightweight helicopter in the world.

— cost-efficient  有成本效益的；合算的

— entry-level  入门的；初级的；最低阶层的；适合于初学者的

译文：R-22 Beta Ⅱ型直升机，以优越的性能、久经考验的安全性，以及实惠的价格和最经济的运营成本，保持在同级别直升机中包括速度、高度、航程的每一项性能记录，成为世界上最受欢迎的入门级轻型直升机。

5. Thanks to the advantageous combination of speed, performance and maneuverability, Bell 407 helicopter has established an outstanding market acceptance. Its spacious passenger cabin can carry 7 passengers and can be flexibly configured within the payload range in accordance with the business requirements.

— market acceptance  市场接受度

— in accordance with  依照；与……一致

译文：贝尔407直升机因速度、表现与操控性的完美结合，打造了卓越的市场认可。宽敞的客舱可搭载7名乘客，并能根据业务需要在有效载荷范围内作灵活配置。

6. Bell 407GX helicopter has the total length at 12.7 meters, fuselage by 2.68 meters wide and 3.56 meters high, cruise speed at 246 km/h, maximum takeoff weight at 2,722 kilograms, maximum climbing altitude at 5,157 meters and maximum range at 611 kilometers, together with 7 seats and a Rolls-Royce 250-C47B/8 engine.

— climbing altitude  爬升高度

— Rolls-Royce  罗尔斯·罗伊斯（又称劳斯莱斯）是英国著名的航空发动机公司，也是欧洲最大的航空发动机企业，它研制的各种航空发动机广为世界民用和军用飞机所采用。罗尔斯·罗伊斯由英文Rolls-Royce翻译而来，也被译为"罗尔斯－罗伊斯"，简称"罗罗"。

译文：贝尔407GX直升机总长度为12.7米，机身宽2.68米，机身高3.56米，巡航速度246公里/小时，最大起飞重量2722千克，最大的爬升高度是5157米，最大航程611千米，座位数7，罗罗250-C47B/8发动机。

## Unit 3  GA aircrafts and manufacture 通用航空飞机及制造

7. In comparison with other models of helicopter of the same class, it enjoys larger payload, longer range and cruise time as well as a cabin so spacious that makes it a high-end model which can load two pairs of stretchers and a single-engine helicopter truly capable of emergency medical service (EMS).

— high-end  高端的；高档的

— emergency medical service (EMS)  紧急医疗服务，也称"急救"。当有任何意外或急病发生时，施救者在医护人员到达前，按医学护理的原则，利用现场适用物资临时及适当地为伤病者进行初步救援及护理，然后从速送院。

译文：在同级别机型中载荷更大，航程和巡航时间更长，宽大的客舱空间使其成为单发直升机中一款能装载两副担架的高端机型，是具有真正紧急医疗服务功能（EMS）的单发直升机。

8. It is low in operational cost, long in time interval of regular inspection, and easy in maintenance. It can satisfy various operational requirements of different customers, and it is mainly used for medical emergency rescue, forest protection and fire prevention, patrol and alarm, satellite hot spot reconnaissance.

— time interval  时段，时间间隔，时距，时间区间

— hot spot  热点；过热点；潜在的危险地区

译文：其运行成本低，定检间隔长，维修简化，可满足不同客户的各种运营要求。主要用于医疗紧急救援、护林防火、巡逻报警、卫星热点侦察。

9. Enstrom 480 helicopter, a lightweight multi-purpose five-seat civil helicopter developed by the USA-based Enstrom Helicopter Corporation, is updated according to the US Army's requirements on the new training helicopter; hence it is reputed for its safety, stability, comfort and good value.

译文：恩斯特龙480（英文：Enstrom 480）直升机，为美国恩斯特龙直升机公司研制的5座轻型多用途民用直升机。它根据美国陆军新训练直升机的要求而改进，所以这架飞机以安全、稳定、舒适、有价值著称。

10. The Enstrom 480B helicopter has the total length at 9.10 meters, diameter of rotor at 9.75 meters, fuselage by 3.0 meters high, cruise speed at 202 km/h, maximum takeoff weight at 1361 kg, practical service ceiling up to 3,962 meters and maximum range at 685 km, together with 5 seats and a Rolls-Royce 250-C20W engine.

— maximum takeoff weight  最大起飞重量是指因设计或运行限制，航空器能够起飞时所容许的最大重量。最大起飞重量是航空器的三种设计重量限制之一，其余两种是最大零燃油重量和最大着陆重量。

— service ceiling  实用升限。飞机能维持平飞的最大飞行高度叫升限，内分理论升限和实用升限。实用升限是在给定的重量和发动机工作状态下，飞机在垂直平面内作等速爬升时，对于亚音速飞行，最大爬升率为0.5/秒时的飞行高度；对于超音速飞行，最大爬升率为5米/秒时的飞行高度。

译文：Enstrom480B直升机总长度为9.10米，旋翼直径9.75米，机身高3.0米，巡航速度202千米/小时，最大起飞重量1361千克，实用升限是3962米，最大航程685千米，座位数5，罗罗250-C20W发动机。

11. It is supposedly accepted as a classic work in the industrial design and

people in its field commented it as the "safest and most reliable" helicopter. Such comment means more its being an industry standard than a mere title of honor. With its widespread global presence, it has exerted far-reaching impact on the development of civil turboshaft helicopter.

— turboshaft engine 涡轮轴发动机。在工作和构造上，涡轮轴发动机同涡轮螺桨发动机很相近，它们都是由涡轮风扇发动机的原理演变而来，只不过后者将风扇变成了螺旋桨，而前者将风扇变成了直升机的旋翼。除此之外，涡轮轴发动机也有自己的特点：它一般装有自由涡轮（即不带动压气机，专为输出功率用的涡轮），而且主要用在直升机和垂直/短距起落飞机上。

译文：这应该算是工业设计中的经典之作，业内称其为"最安全、最可靠"的直升机，这样评价早已超越了荣誉而成为标准。由于它广泛的全球存在性，它深远地影响了民用涡轴直升机的发展。

## Exercises

Section one: Answer the following questions according to the text.

1. What is the generally accepted definition of general aviation mentioned in the paragraph one?
2. How many models of general aviation aircrafts are most commonly used in China?
3. Please make a comparison between Robinson R-44 and H125[AS350B3e] Squirrel, Airbus Helicopter.
4. What are the main application of Robinson R-22?
5. What are the differences in parameter data between Bell 407GX and Agusta AW119KX?

Section two: Match the words or phrases in column A with the definition in column B.

| A | B |
| --- | --- |
| 1. engine | a. the central body of an airplane designed to accommodate passengers |
| 2. cargo | b. a detachment used for security or reconnaissance |
| 3. training | c. a detailed critical inspection |
| 4. fuselage | d. motor that converts thermal energy to mechanical work |
| 5. endurance | e. goods carried by a large vehicle |
| 6. patrol | f. distributed over a considerable extent |
| 7. standard | g. activity leading to skilled behavior |
| 8. survey | h. the power to withstand hardship or stress |
| 9. widespread | i. ahead in development; complex or intricate |
| 10. sophisticated | j. a basis for comparison |

## Section three: Translate the following into Chinese.

1. By international practice, general aviation refers to all aircrafts, flight scope and activities excluding air force and passenger and cargo transportation of civil aviation.

2. There're now 200 such helicopters in China which are mainly applied to pilot training program, private flight, commercial flight and so on.

3. The H125 [AS350B3e] helicopter is a six-seat single-engine helicopter featuring solid and durable performance, high reliability and low cost.

4. The H125 helicopter is 12.94 meters long, 2.53 meters wide, 3.34 meters high, with cruise speed at 251 km/h, maximum weight at 2,800 kg, maximum climb altitude at 5,258 meters and maximum range at 686 km.

5. The R22 Beta II helicopter, by the strength of its superior performance, proven safety, affordable price and the most cost-efficient operation, keeps every record in the performance of speed, altitude and flight range among helicopters of same class.

6. Its multiple power-driven components and outperformed flight control can ensure a comfortable piloting and boarding experience in various environments.

7. With the flexible operation features, it can be applied widely in every field of general aviation, and be reconfigured quickly according to the requirements of missions.

## Section four: Translate the following into English.

1. 任何涉嫌与占领军勾结的人都被逮捕了。
2. 每个工会选出数名代表参加年度大会。
3. 财政决策权已经交给一个特别委员会了。
4. 一个由4名教师组成的小组被选出来代表同事出席工会大会。
5. 请不要打扰你姐姐——她正在做家庭作业。
6. 几名球迷被捕,他们被指控在赛后扰乱治安。
7. 某些场景中有暴力内容,可能会使年幼些的观众产生焦虑情绪。
8. 她说自己在工作中受到的待遇让她在情感和心理上都感到非常痛苦。
9. 直升机从海岸附近的一艘遇险渔船上救出了6个人。
10. 他试图转移人们对其非法活动的注意。

# 3.2 Major manufacturers of GA aircraft
## 主要通用航空飞机制造商

## Background

According to the current market value of general-purpose aircraft products, manufacturers with the highest share are respectively Bombardier Aerospace Group, Gulfstream Aerospace Corporation, Cessna Aircraft Company, Hawker Beechcraft Corporation (formerly Raytheon Company) and Dassault Aircraft Company, and their aircraft products usually include various types, such as business aircraft, multi-purpose aircraft, etc. As to civil helicopter companies, Bell Helicopter, Textron, Eurocopter and Agust Westland comes out on top with the highest global market share.

### 1. Bombardier Aerospace Group

Bombardier Inc., the parent company of Bombardier Aerospace Group, is a private company established in 1942 to produce snow transportation vehicles. By 1974, the company had grown into a large industrial enterprise. In 1986, the company made inroads into the aerospace sector and acquired Canada Air, Canada's largest aircraft manufacturer. In 1988, the company expanded in Europe and became a major shareholder of Belgium-based BN Company, a transport equipment company with advanced technology. A year later, the company continued to take over Short Brothers PLC in the UK and ANF railway vehicle manufacturing plant in France, and in 1990, the company bought American Learjet Corporation, and in 1992 De Haviland Inc. in Toronto, Canada, from Boeing. As a powerful industrial group, Bombardier restructured its business in 2003 to focus on aerospace and transportation sector including railway transportation equipment, regional aircraft and business aircraft. Now the company consists of three major business groups, namely Bombardier Aerospace Group, Bombardier Transportation Group and Bombardier Capital Group. As of January 31, 2006, Bombardier employed a total of 55800 people, and in fiscal year of 2006, its operating revenue reached 14.7 billion US dollars.

As one of the business groups under Bombardier Inc., Bombardier Aerospace Group is headquartered in Santville near Montreal International Airport, Canada. The Group has four aircraft manufacturing enterprises, i.e. Canadian Aerospace Company in Saint Laurent, Quebec, Canada, De Haviland Inc in Downsville, Ontario, Canada,

Gates Lear Jet Company in Tucson, Arizona, USA, and Short Brothers Company based in Belfast, Northern Ireland. Over the past 15 years, Bombardier Aerospace Group has stridden forward into the third largest civil aircraft manufacturer in the world, and the Group, with its flexible marketing and management strategy, is also a manufacturer complete with full capabilities of design, manufacture and sale of aviation products in both Canada, the United States and the United Kingdom, the three countries where the Group's four aviation manufacturing enterprises are located.

Bombardier Aerospace Group now includes five main business departments, namely business aircraft department, regional aircraft department, aircraft service and new commercial aircraft project department, amphibious aircraft department as well as Flexjet and Skyjet Company. In the Bombardier business aircraft department, its main products are narrow-body and wide-body business jet, including Learjet, Challenger, and Global Express business aircraft. Regional aircraft department undertakes products such as CIU regional jet and Q200, Q300, Q400 turboprop aircraft. Aircraft service and new commercial aircraft project department mainly render services of parts support, aircraft maintenance, customer training, military aviation training, governmental and mission-based aircraft. Amphibious aircraft department is highlighted with its key product Bombardier 415. And as to the Flexjet and Skyjet Company, it provides partial aircraft ownership service and hourly flight permission programs. The Flexjet company has received 156 Bombardier business aircrafts in the past decade, and the Skejet company provides charter service in various forms either through Skyjet or Skyjet International. The Skyjet presents in North American region, whilst the Skyjet International operates businesses in Europe, Asia Pacific and the Middle East with regional offices based in London, Hong Kong and Dubai. As of fiscal year ended January 31, 2006, Bombardier Aerospace Group had a total of about 26800 employees, generated $8.1 billion in revenue and delivered 337 aircrafts, including 186 business aircrafts (partial ownership projects included) and 2 amphibious aircrafts.

## 2. Gulfstream Aerospace Corporation

Headquartered in Savannah, Georgia, USA, Gulfstream is a world famous manufacturer of luxury and large business aircraft. In 1958, Gulfstream I was launched as the first business aircraft designed exclusively for business applications. In 1966, Gulfstream II founded the market of large cabin business aircraft. And in 1978, Alan Paulson purchased the production line of Gulfstream aircraft from Grumman for $2 million and took over Gulfstream's various programs then. Gulfstream was born. In 1990, Paulson joined Forstmann Little & Co and sold Gulfstream to the company for $850m, in which he held 32% shares of Gulfstream and remained the president of Gulfstream. In 1980, with the advent of Canadair's Challenger and the subsequent expansion of Bombardier Aerospace, the market of Gulfstream company was eroded. In 1992, Gulfstream company started the development of Gulfstream V, an ultra long-range business jet, which is not easy to reduce costs. The jet was proved to be a commercial success and was produced more than one thousand. In 1999, General Dynamics purchased Gulfstream from Forstmann Little & Co, and in 2001, it merged Galaxy Aerospace, which was purchased from IAI, into Gulfstream.

General-purpose aircraft products
8 current models of business aircraft manufactured by Gulfstream
G100/G150 : Based on IAI Astra SPX
G200 : Based on IAI Galaxy
G300/G350 : Based on Gulfstream IV-SP
G400/G450 : Based on Gulfstream IV-X
G500/G550 (see the photo below) : Based on Gulfstream V-SP(launched in 2002)

### 3. Cessna Aircraft Co.

Cessna Aircraft Company, headquartered in Wichita, Kansas, USA and a subsidiary of Textron, is the world's largest private aircraft manufacturer. The company was founded in 1927 by Clyde Cessna and initially made monoplane, but from 1932 to 1934, it shut down due to the economic downturn in the United States. In 1934, Dwane Wallace, under the assistance of his brother Dwight, took over the company and made it grow in steps, and the company came through with flying colors in the small general aircraft market all over the world. After the Second World War, Cessna Aircraft Company rolled out the Cessna 170 aircraft, which became the most produced lightweight aircraft in history.

In 1985, Cessna Aircraft was acquired by General Dynamics. In 1986, the company stopped the production of single engine piston aircraft. In 1992, Textron purchased Cessna Aircraft Company and resumed the production of single-engine lightweight piston aircraft in 1994.

The main products of Cessna Aircraft Company are Citation business aircraft, Caravan single engine multi-purpose turboprop aircraft, and the improved single engine propeller-induced piston aircraft No. 172, 182, 206. Cessna also provides partial ownership program for its business aircraft. In 2006, Cessna Aircraft accounted for 36% of Textron's sales.

### 4. Bell Helicopter Textron

Bell Helicopter Textron is a famous civil and military helicopter and tilt-rotor aircraft manufacturer in the United States. Headquartered in Fort Worth, Texas, it is now a subsidiary of Textron and also known as the world's largest helicopter manufacturer.

Formerly founded by Larry Bell in 1935 as Bell Aircraft Company, the company started its business mainly by designing and manufacturing fighters. In 1941, the company began to set foot in helicopter business and launched its first helicopter, Bell 30. In 1960, Textron acquired Bell Aerospace, including Bell's helicopter division, which later was renamed as Bell Helicopter Company. A few years later, the company's UH-1 helicopter was widely used. In January 1976, the name of Bell Helicopter Company was changed to Bell Helicopter Textron.

Bell's civil helicopter can offer a ride up to 15 people and is designed for transportation, emergency medical services, search and rescue activities. Bell also works with Agust Westland to manufacture helicopters and provide helicopter maintenance, repair and overhaul services. Bell Helicopter Textron mainly produces military-purpose helicopter and tilt-rotor aircraft in Fort Worth and Amarillo, Texas, and civil helicopters in Mirabel, Quebec, Canada. In 2006, Bell Helicopter Textron sold 109 civil helicopters, or 13.87% of the total world sales. In fiscal year of 2007, the company's sales revenue was 599.2 million US dollars, with more than 8000 employees.

## General Vocabulary

Bombardier Aerospace Group　庞巴迪宇航集团
Gulfstream Aerospace Corporation　湾流公司
Cessna Aircraft　Company　赛斯纳飞机公司
Hawker Beechcraft Corporation (formerly Raytheon Company)　豪客比奇飞机公司（原雷神公司）
Dassault Aircraft Company　达索公司
Bell Helicopter　贝尔直升机
Textron　德事隆公司
Eurocopter and Agust Westland　欧洲直升机公司和阿古斯特·韦斯特兰公司
inroad　n. 损害；得手；侵犯；消耗
acquire　vt. 获得；取得；学到；捕获
shareholder　n. 股东；股票持有人

belgium　*n.* 比利时（西欧国家，首都布鲁塞尔 Brussels）
take over　接管；接收
railway　*n.*〔英〕铁路；轨道；铁道部门
Learjet Corporation　里尔公司
De Haviland Inc.　德·哈维兰飞机公司
Toronto　*n.* 多伦多（加拿大城市）
Boeing　波音公司
aerospace　*n.* 航空宇宙；航空航天空间
fiscal year　会计年度；〔美〕财政年度
Montreal　*n.* 蒙特利尔（加拿大东南部港市）
Saint Laurent　圣洛朗
stride　*vt.* 跨过；大踏步走过；跨坐在……
the United Kingdom　英国
amphibious　*adj.* 两栖的，水陆两用的；具有双重性的
narrow-body　窄体
wide-body　宽体
turboprop　*n.* 涡轮螺桨发动机；涡轮螺旋桨飞机
highlight　*v.* 突出；强调；使显著；加亮；着亮彩于（头发）
partial　*adj.* 局部的；偏爱的；不公平的
permission　*n.* 允许，许可
whilst　*conj.* 当……的时候；与……同时；然而；虽然，尽管；直到……为止（与while同）
regional office　地区办事处，办事处，地区办公室
Dubai　*n.* 迪拜
luxury　*n.* 奢侈，奢华；奢侈品；享受
exclusively　*adv.* 唯一地，专有地，排外地；作为唯一的（消息）来源
advent　*n.* 到来；出现；基督降临；基督降临节
subsequent　*adj.* 随后的
expansion　*n.* 膨胀；阐述；扩张物
erode　*vt.* 腐蚀，侵蚀
Kansas　*n.* 堪萨斯州（美国州名）
ultra　*adj.* 极端的，偏激的
long-range　*adj.*（飞机、火箭等）远程的；长期的；远大的
merge　*vt.* 合并；使合并；吞没
galaxy　*n.* 银河；星系；银河系；一群显赫的人
subsidiary　*n.* 子公司；辅助者
monoplane　*n.* 单翼机
downturn　*n.* 衰退（经济方面）；低迷时期
flying colors　成功；飘扬的旗帜；胜利
roll out　推出，铺开，转出

piston　n. 活塞

resume　vt.（中断后）重新开始，继续；重新回到，恢复（席位，地位或职位）；（停顿或被打断后）继续说；重选，重穿，重新占用

citation　n. 引用，引证；传票；褒扬

propeller　n. 螺旋桨；推进器

tilt　v. 倾斜，翘起；俯仰（摄影机）；以言词或文字抨击

set foot in　踏进；进入；涉足于

division　n. 除法；部门；分配；分割；师（军队）；赛区

rename　vt. 重新命名，改名

## Notes

1. A year later, the company continued to take over Short Brothers PLC in the UK and ANF railway vehicle manufacturing plant in France, and in 1990, the company bought American Learjet Corporation, and in 1992 De Haviland Inc. in Toronto, Canada, from Boeing.

— Short Brothers PLC　是一家英国老牌的航空工业企业，通常简称肖特（Shorts），总部位于北爱尔兰的贝尔法斯特，是北爱尔兰最大的制造企业，成立于1908年，是世界上第一家真正的飞机制造公司。1911年，肖特兄弟申请了多台发动机飞机设计的专利，这在航空史上是一个重大进步。

— De Haviland Inc.　德哈维兰公司是一家英国飞机、飞机发动机制造商。公司已于1964年合并至霍克薛利公司。德哈维兰公司由著名飞机设计师杰弗里·德·哈维创立，在第二次世界大战中开发出了赫赫有名的蚊式轰炸机，还在1949年研发出了世界上第一款喷气式客机彗星飞机。1927年，德哈维兰在澳大利亚成立了分公司de Havilland Aircraft Pty.Ltd（简称DHA），该公司后来被波音公司澳大利亚分公司收购。德哈维兰于1928年在加拿大成立德哈维兰加拿大公司（简称DHC），该公司开发冲7（Dash7）、冲8（Dash8）等成功机型，之后于1986年被庞巴迪公司收购。

译文：1989年，该公司收购了英国肖特兄弟公司(Short Brothers PLC)和法国ANF铁道车辆制造厂。1990年，该公司收购了美国利尔喷气机公司(Learjet Corporation)。1992年，该公司从波音公司手中收购了位于加拿大多伦多市的德哈维兰飞机公司。

2. Over the past 15 years, Bombardier Aerospace Group has stridden forward into the third largest civil aircraft manufacturer in the world, and the Group, with its flexible marketing and management strategy, is also a manufacturer complete with full capabilities of design, manufacture and sale of aviation products in both Canada, the United States and the United Kingdom, the three countries where the Group's four aviation manufacturing enterprises are located.

— stride foreward　跨过；大踏步走过；跨坐在……

译文：15年来，庞巴迪宇航集团已发展成为世界第三大民用飞机制造商，该集团所属的四家航空制造企业分布在加拿大、美国和英国，因此，该集团是在三个国家内具有完备的设计、制造、销售航空产品能力的制造商，市场营销和管理都相当灵活。

3. Bombardier Aerospace Group now includes five main business departments, namely business aircraft department, regional aircraft department, aircraft service and new commercial aircraft project department, amphibious aircraft department as well as Flexjet and Skyjet Company.

- regional aircraft  支线飞机
- amphibious aircraft  两栖飞机。"水陆两栖飞机"是指能在水面上起飞、降落和停泊的飞机，可以适应水上、空中两种不同环境。乘坐水上飞机的乘客可以获得和大飞机不一样的体验，飞行员可以根据乘客需求做一些转弯等大动作。它适用于海上巡逻、反潜、救援和森林灭火等任务。现代水陆两用飞机在很多应用领域里已被直升机所取代，但是它在森林或城市灭火方面仍有独特的作用。

译文：目前，庞巴迪宇航集团主要分为庞巴迪公务机部、庞巴迪支线飞机部、飞机服务和新商用飞机项目部、庞巴迪两栖飞机部以及庞巴迪Flexjet和Skyjet公司5个业务部门。

4. In the Bombardier business aircraft department, its main products are narrow-body and wide-body business jet, including Learjet, Challenger, Global Express business aircraft. Regional aircraft department undertakes products such as CIU regional jet and Q200, Q300, Q400 turboprop aircraft.

- narrow-body  窄体机，窄体客机是每排座位不超过六座的单走道客机，属于客机的一种。
- wide-body  宽体机，外直径5到6米（16到20英尺），并且有两条走道，通常一排能够容纳7到10个座位。
- Challenger  挑战者，Challenger（挑战者）系列私人飞机由著名的飞机制造商加拿大Bombardier Aerospace（庞巴迪宇航公司）制造，作为全球最大型的商务机生产商，Challenger系列代表了庞巴迪公司最高的技术结晶，其中以Challenger 850系列最为闻名。
- Global Express  环球系列，"环球5000"是一种超巨型商用飞机，与同级别的喷气式公务机相比，庞巴迪环球5000在飞机性能以及机舱的舒适性上均有其独到之处。这一款动力强劲、内饰奢华的超远程公务机最多可以乘坐17名乘客。环球6000继承了"环球快车"系列的优良品质，迎合了国内高端客户最为挑剔和苛刻的要求，它宽敞的机舱空间和出色的远程飞行能力特别受频繁往返于各个大洲的商界领袖们青睐。
- turboprop  n. 涡轮螺桨发动机；涡轮螺旋桨飞机。是一种通常用于飞机上的燃气涡轮发动机（gas turbine engine）。涡桨发动机的驱动原理大致上与使用活塞发动机作为动力来源的传统螺旋桨飞机雷同，是以螺旋桨旋转时所产生的力量来作为飞机前进的推进力。其与活塞式螺旋桨飞机主要的差异点除了驱动螺旋桨中心轴的动力来源不同外，还有就是涡桨发动机的螺旋桨通常是以恒定的速率运转，而活塞动力的螺旋桨则会依照发动机的转速不同而有转速高低的变化。代表有俄罗斯HK-93函道桨扇发动机。

译文：庞巴迪公务机部的主要产品是窄体和宽体公务机，包括"利尔喷气"系列公务机和"挑战者"系列公务机以及庞巴迪"全球"系列公务机。支线飞机部的产品包括支线喷气机CIU系列以及涡桨飞机Q200、Q300、Q400。

5. Amphibious aircraft department is highlighted with its key product Bombardier 415. And as to the Flexjet and Skyjet Company, it provides partial aircraft ownership service and hourly flight permission programs.

— Bombardier 415　CL-415飞机是加拿大庞巴迪宇航公司研制的水陆两栖双发螺旋桨飞机。CL-415是在CL-215的基础上开发出来专门用于扑救森林火灾的机型，其灭火作战能力更为强大。CL-415执行任务范围很宽，适合在陆地和水域使用。按照设计，CL-415是一架非常高效的灭火飞机，可以从着火地区的小河流、湖泊、海岸或海上汲水，在距水源6-15千米距离内，仅需5-12分钟便可周转一次，灭火效果显著。

译文：两栖飞机部主要产品为庞巴迪415。Flexjet和Skyjet公司提供飞机部分拥有服务和按小时飞行权限项目。

6. In 1980, with the advent of Canadair's Challenger and the subsequent expansion of Bombardier Aerospace, the market of Gulfstream company was eroded. In 1992, Gulfstream company started the development of Gulfstream Ⅴ, an ultra long-range business jet, which is not easy to reduce costs. The jet was proved to be a commercial success and was produced by more than one thousand jets.

— an ultra long-range business jet　超远程喷气式公务机

译文：1980年加拿大飞机公司"挑战者"的问世和之后庞巴迪宇航公司的壮大，使湾流公司的市场被侵蚀。湾流公司在不易降低成本的情况下，于1992年开始了超远程喷气公务机"湾流"Ⅴ的研制，该机获得了商业成功，生产了千余架。

7. G500/G550：Based on Gulfstream V-SP(launched in 2002)

湾流G550公务机（Gulfstream G550），由美国湾流宇航公司于2003年推出，它是人类飞行史上首架直航范围能从纽约直达东京的超远程公务飞机。湾流G550公务机是国际顶级远程喷气式公务机代表机型之一，航程为11686公里，最大巡航高度为15545米，可搭载18名乘客，是国内航程最远、性能最优、客舱最宽敞、舒适性最好的豪华公务机。

8. In 1934, Dwane Wallace, under the assistance of his brother Dwight, took over the company and made it grow in steps, and the company came through with flying colors in the small general aircraft market all over the world. After the Second World War, Cessna Aircraft Company rolled out the Cessna 170 aircraft, which became the most produced lightweight aircraft in history.

— take over　接收，接管

— flying colors　成功；飘扬的旗帜；胜利

— lightweight aircraft　轻型飞机，一般是指最大起飞重量小于5700千克的飞机。这种飞机具有轻便、安全、使用要求低、能在草地起降、易于操作、价格低廉等特点。轻型飞机体积小，构造相对简单，很适于休闲和商业旅行。尽管一些轻型飞机装有喷气发动机，但通常情况下都使用活塞发动机驱动推进器。大型高速飞机装有起落架，可在起飞后将轮子收起。轻型飞机为减轻重量和节约费用，通常装备固定脚架。随着新型合成材料的开发，轻型飞机将变得更坚固，更轻巧，并且可以飞得更远。

译文：1934年，德万·华莱士（Dwane Wallace）在其兄弟德怀特的帮助下接管了该公司，使其逐步发展壮大，在全世界小型通用飞机市场取得了成功。第二次世界大战后，赛斯纳飞机公司推出了赛斯纳170飞机，该机成为历史上产量最大的轻型飞机。

9. In 1992, Textron purchased Cessna Aircraft Company and resumed the production of single-engine lightweight piston aircraft in 1994.

— piston engine　活塞式发动机。从1903年世界第一架飞机到第二次世界大战末期，所有飞机都用活塞式航空发动机作为动力装置。20世纪40年代中期以后，在军用飞机和大型民用飞机上燃气涡轮发动机逐步取代了活塞式航空发动机，但小功率活塞式航空发动机比燃气涡轮发动机经济，在轻型低速飞机上仍得到应用。

译文：1992年，德事隆公司购买了赛斯纳飞机公司，并于1994年恢复了单发活塞轻型飞机的生产。

10. The main products of Cessna Aircraft Company are Citation business aircraft, Caravan single engine multi-purpose turboprop aircraft, and the improved single engine propeller-induced piston aircraft No. 172, 182, 206.

— Citation business aircraft "奖状"系列公务机。1996年，赛斯纳新一代入门级中型公务机"奖状Excel"（优胜号Model 560 XL）获得FAA型号适航认证并在1997年开始交付。在客舱高度可站立（1.73米）、最大起飞重量超过2万磅（9072千克）的中型喷气公务机中，"奖状优胜"号购买和使用成本最低，舒适而经济，1800多海里的航程，每小时400多海里的巡航速度，45000英尺的升限，对于大多数经常作中短航程（2500千米以内）旅行的企业家们来说，它已经完全能满足使用需要了。因此"奖状优胜"号一经推出，即深受市场喜爱，连续多年供不应求。

译文：赛斯纳飞机公司主要产品有"奖状"系列公务机、"大篷车"系列单发涡桨多用途飞机以及改进后重返市场的172、182、206活塞单发螺旋桨飞机。另外赛斯纳还为其公务机提供部分拥有项目。

11. In 1960, Textron acquired Bell Aerospace, including Bell's helicopter division, which later was renamed as Bell Helicopter Company. A few years later, the company's UH-1 helicopter was widely used.

— UH-1是美国的贝尔直升机公司所设计制造的军用中型通用直升机，UH-1直升机为多用途设计，其U代表通用（Utility），从运补作业到攻击任务皆宜，是美国批量装备的第一个搭载了涡轮轴发动机的直升机。它的民用版本是Bell 204。

译文：1960年，德事隆公司收购了贝尔航宇公司，其中包括贝尔的直升机分部，随后该分部更名为贝尔直升机公司。几年后公司的UH-1直升机获得广泛应用。

12. Bell's civil helicopter can offer a ride up to 15 people and is designed for transportation, emergency medical services, search and rescue activities. Bell also works with Agust Westland to manufacture helicopters and provide helicopter maintenance, repair and overhaul services.

— Agust Westland 阿古斯塔·韦斯特兰公司商用直升机迅速地在多种用途领域取得了成功。阿古斯塔·韦斯特兰公司直升机可执行执法、减灾救援、海事巡逻、要员（VIP）运输和客运、海港往返领航等多种任务。阿古斯塔·韦斯特兰公司的直升机尤其在执法和减灾救援任务方面取得了巨大的成功。

译文：贝尔的民用直升机可乘坐多达15人，其直升机设计用于运输、紧急医疗服务以及搜索与救援活动。贝尔公司还与阿古斯特·韦斯特兰公司合作制造直升机，以及提供直升机修理、维修与大修服务。

# Exercises

## Section one: Answer the following questions according to the text.

1. List some general aviation aircraft manufacturers with high market shares.
2. What are the main business departments in Bombardier Aerospace Group?
3. Please briefly describe the development history of Gulfstream Aerospace Corporation.
4. What are the typical general-purpose aircraft products of Gulfstream Aerospace Corporation?
5. How did the second World War affect the progress of Cessna Aircraft Company rolling out the Cessna 170 aircraft?

## Section two: Match the words or phrases in column A with the definition in column B.

| A | B |
| --- | --- |
| 1. product | a. the acquisition of something for payment |
| 2. maintenance | b. join or combine |
| 3. remain | c. approval to do something |
| 4. cabin | d. connected with or engaged in commerce |
| 5. merge | e. commodities offered for sale |
| 6. purchase | f. characteristic of a region |
| 7. regional | g. small room on a ship or boat where people sleep |
| 8. permission | h. a specialized division of a large organization |
| 9. commercial | i. stay the same; remain in a certain state |
| 10. department | j. activity involved in maintaining something in good working order |

## Section three: Translate the following into Chinese.

1. Bombardier Aerospace Group now includes five main business departments, namely business aircraft department, regional aircraft department, aircraft service and new commercial aircraft project department, amphibious aircraft department as well as Flexjet and Skyjet Company.

2. As of fiscal year ended January 31, 2006, Bombardier Aerospace Group had a total of about 26800 employees, generated $8.1 billion in revenue and delivered 337 aircrafts, including 186 business aircrafts (partial ownership projects included) and 2 amphibious aircrafts.

3. In 1992, Gulfstream company started the development of Gulfstream V, an ultra long-range business jet, which is not easy to reduce costs. The jet was proved to be a commercial success and was produced more than one thousand.

4. In 1934, Dwane Wallace, under the assistance of his brother Dwight, took over the company and made it grow in steps, and the company came through with flying colors in the small general aircraft market all over the world.

5. Bell Helicopter Textron is a famous civil and military helicopter and tilt-rotor aircraft manufacturer in the United States.

6. In 2006, Bell Helicopter Textron sold 109 civil helicopters, or 13.87% of the total world sales. In fiscal year of 2007, the company's sales revenue was 599.2 million US dollars, with more than 8000 employees.

Section four: Translate the following into English.
1. 我觉得灰白的头发使男人看起来很睿智。
2. 我最初的说法被媒体彻底歪曲了。
3. 她的嗓音很独特。
4. 工会要求今年加薪7%。
5. 他总是以最高的行为准则来要求自己的孩子。
6. 他似乎缺少成功政治家所应具备的很多素质。
7. 新工作对他要求很高。
8. 部门内部的各项职责都有明确的界定。
9. 停车位用白线标了出来。
10. 公司财务紧张使得他们不能再聘请新员工。

# 3.3 Business aircraft
# 公务机

## Background

### 1. On-demand air transport

All means of transportation can be grouped into either regular transport or non-regular transport. The regular transport means that the transportation is operated in accordance with the confirmed timetable and its time and itinerary is expressly indicated without any need for particular explanation, such as airway transport, railway transport, highway transport, waterway transport, etc. The non-regular transport, instead, calls for further explanation and clarification as the word non-regular itself embodies randomness and irregularity.

Hence, to define what non-regular transport is, the general concepts will fail to present its complete and exact connotations. Likewise, the concept of on-demand air transport requires more detailing, though it refers that certain air transport service can be in place to satisfy the needs and requirements raised by customers. This section is dedicated to elaborating various types of on-demand air transport and also imparting to readers how to choose and make use of them.

On-demand air transport brings advantages and benefits to its variety of users including individuals, company managers, troubleshooters, sport teams, entrepreneurs and some families. They share one same goal that they could reach a certain destination as safely and comfortably as possible. Meanwhile, there're some individualized reasons. Some take it as a sign of their social status; some take more account of its economical benefit; some prefer a faster or recreational travel; some thrill-seekers challenge to pilot themselves, while others would prefer to outsource the flight missions to professional flight crew. In any case, their purpose is to keep to their own schedule, so as to master and control their own pace of life more actively.

As personal aviation could be operated by the aircraft owner himself or provided by commercial enterprises when called upon, thus, it usually boils down to a part of commercial aviation or general aviation.

Business aviation encompasses all types of air flights for the purpose of business activities, under which the main types are employee-flown operations and corporate operations. Employee-flown operation is primary that the aircraft owner himself operates the aircraft and the aircraft could be piloted by any of his employees to transport other employees or customers. Hence the key of employee-flown operation relies on that the aircraft is operated by the company's one or several own pilots without a special need to employ professional pilots or pay particular expenses. In stark contrast, corporate operations deploy professional pilots (usually 2 pilots) for commercial end and shall effect payment for the flight service undertaken by the pilots. NBAA (National Business Aviation Association) defines corporate aviation as follows: to promote business, a company or enterprise undertakes to operate self-owned or leased aircrafts to carry staff or commodities by employing professional pilots who will directly receive salary or special rewards on flight task basis. It is also noted that the companies, being employee-flown operations or corporate operations, do not have to own or contract long-term aircraft leasing, and instead, they can rent or borrow aircraft respectively (though some restrictions may apply).

In other way, using charter companies and partially owned aircraft can also meet the needs of commercial enterprises for on-demand air transport.

### 2. Business aviation

Every year thousands of people travel around the world for business activities so as to establish close ties, exchange ideas, seize business opportunities and achieve face-to-face communication. While business intercourse is crowded with phone calls, emails, faxes and postal mails, people are also in need of direct in-person communications to finalize business contracts, address thorny issues and reach new consensus.

According to the prediction of technical experts, business travel will plunge due to the emergence of mobile phones, networks and teleconferencing. Their theoretical research points out that better and newer communication equipment that weakens direct in-person talks and teleconference will rise as a best alternative. However, when these new telecommunication equipment should arguably cut some business travels, the odd fact is that both the teleconference and business travel burgeon. As a matter of fact, as new communication technology increases opportunities and enhances productive efficiency for businesses, business travel in turn also increases. Thanks to the two correlated ways of communication, countries including the United States, Germany, and Japan are in the forefront of the world's production efficiency.

In business circles, the demand for face-to-face negotiation has never faltered but being growing. New important trade is often done in person and rarely by phone or Internet. Through meetings, we can build up friendship with customers, understand their needs and feedbacks, help them clear up problems, and introduce new products and services. Business travel is a long-term industry.

Every year, airlines around the world carry 1 billion domestic passengers and 550 million international passengers out of which nearly one third travel for business purposes. Through thousands of airports, airlines provide frequent, safe and reliable services for global business travelers, and around two-thirds of business travels were completed by such airlines. Nevertheless, there're also other means of travel.

Taking into account of such many airlines already available to transport passengers to any part of the globe, some observers and commentators, when it comes to business aviation, may pose questions like why do we still need another aircrafts to transport a small number of passengers to the same destination? Isn't it expensive? And how safe is it?

In 2002, more than 13000 operators deployed a total of above 21000 gas turbine (turbojet and turboprop) aircrafts to the business aircraft market. And three quarters of these operators and aircrafts were based in the North America. In the United States alone, there were about 14000 gas turbine business aircrafts with an average in-flight time of 450h per aircraft and approximately 18 million business passengers served in the year of 2001. But interestingly, four fifths of American business aircraft operators own one aircraft only.

**Numbers of business aircrafts in 2001**

| | |
|---|---|
| The United States | 14079 |
| Brazil | 667 |
| Canada | 656 |
| Mexico | 540 |
| France | 431 |
| Germany | 378 |
| Venezuela | 288 |
| The United Kingdom | 283 |
| South Africa | 247 |
| Argentina | 216 |
| Australia | 213 |

Source of data: International Business Aviation Council (IBAC)

Of the Fortune 500 companies, only 29% don't have their own business aircrafts. And the companies in possession of business aircrafts greatly outperform those without the aircrafts in terms of their productivity, sales revenue and profits. The companies take business aircraft as a tool to improve productivity, through which they can take control of the tight travel schedule and facilitate the daily traffic of their business managers who need to attend every production activity. In fact, business aircraft turns out to be as much as the modernized magic time machine in the Arabian Nights, bringing managerial personnel safely and quickly to the next business opportunity.

On-demand air transport has become a constituent part of corporate culture, which enables business activities to be implemented in the foreseeable future. In comparison, those enterprises owning business aircrafts are palpably in better position than their rivals, because air transport by business aircrafts makes business operation characteristically time-saving, flexible, highly efficient, privacy-friendly and safe, and in the same time naturally attaches intangible benefits, such as longer relaxing time at home for travelers, and more comfortable non-stop or minimum-transit travel to the farthest destination. Advantages as listed above are powerful enough to drive the widespread users to like better such mean of transport.

There's no doubt that, to the contentment of board of directors and shareholders, on-demand air transport is the necessary way leading to higher productivity. Business aircrafts, virtually a time magnifier, can take off and land at up to 5000 airports in the United States (Airlines serve only at 500 airports) and these turbojets also enjoy a unparalleled speed and convenience compared with airlines. If a business aircraft can make it possible that the company's key personnel are carried to a remote place to handle business there for half a day and they are sent home for dinner on the same day, then there is no reason to do the same thing in two days. It is an obvious productivity enhancement effect that exactly makes business jet transportation marketable every day.

Moreover, the accounting departments are also awake to the profitability of business aircrafts. Research shows that companies that own airplanes, whether in terms of sales volume, earnings per share, long-term returns to investors, or labor productivity (sales per employee), are growing faster than those that don't have corporate jets. There're already no room for doubt over the fact that the business jet is beneficial to the basic interests of the enterprise. In addition, the current situation of airlines is constantly driving more and more business managers prone to business aviation as well. The regular airline transport is now forced to restrict and divide its own plans and routes to be more competitive, and because of the frequent overbooking, passengers in result cannot take flights and the terminal management is also reduced to be out of control. All of these are calling for more important role of business jet operation, or in other words, the status quo of the airlines may be the best reason for travelers to shift to business jets.

Global village is gradually transformed to an international market in which the application of business aircraft will greatly facilitate such transition. As a matter of fact, the intercontinental business aircraft has already been put into operation for many years, and will continue to create more benefits. Among which, the most popular business aircraft is the turbojet which can fly from New York to Tokyo and from Los Angeles to Paris.

However, corporate aviation is not immune from business cycle as the sound development of corporate aviation department is directly linked to the financial situation of the enterprise. When the business is booming, there will be more flight activities; when the business is depressed, not only will there be less flight, but also there will be less flight time and personnel. The flight will present ups and downs with the trend of the financial situation, but the operation of business aircraft services will sustain and grow even more powerful with the deeper integration of on-demand aviation operation into the corporate culture.

## General Vocabulary

timetable  *n.* 时间表；时刻表；课程表
itinerary  *n.* 旅程，路线；旅行日程
expressly  *adv.* 清楚地，明显地；特别地，专门地
indicate  *vt.* 表明；指出；预示；象征
waterway  *n.* 航道；水路；排水沟

clarification  *n.* 澄清，说明；净化
embody  *vt.* 体现，使具体化；具体表达
randomness  *n.* 随意；无安排；不可测性
irregularity  *n.* 不规则；无规律；不整齐
connotation  *n.* 内涵；含蓄；暗示，隐含意义；储蓄的东西（词、语等）
likewise  *adv.* 同样地；也
elaborate  *vt.* 精心制作；详细阐述；从简单成分合成（复杂有机物）
impart  *vt.* 给予（尤指抽象事物），传授；告知，透露
troubleshooter  *n.* 解决纠纷者；故障检修工；（计算机）故障查找软件
entrepreneur  *n.* 企业家；承包人；主办者
economical  *adj.* 经济的；节约的；合算的
thrill-seeker  寻找惊险刺激的人
outsource  *vt.* 把……外包
call upon  号召；要求；拜访
boil down to  归结起来是；其结果是；归结为；归根结底
encompass  *vt.* 包含；包围，环绕；完成
in stark contrast  形成了鲜明的对比
deploy  *vt.* 配置；展开；使疏开
effect  *vt.* 产生；达到目的
NBAA (National Business Aviation Association)  *abbr.* 美国公务航空协会
leased  *v.* 出租；租用；租借（lease 的过去式和过去分词）
commodity  *n.* 商品，货物；日用品
partially  *adv.* 部分地；偏袒地
seize  *vt.* 抓住；夺取；理解；逮捕
exchange  *vt.* 交换；交易；兑换
face-to-face  *adj.* 面对面的；当面的
in-person  亲自；外貌上
finalize  *vt.* 完成；使结束
address  *v.* 写（收信人）姓名地址；演说；向……说话，致辞；冠以（某种称呼）；设法解决；就位击（球）；提出
thorny  *adj.* 多刺的；痛苦的；令人苦恼的
prediction  *n.* 预报；预言
plunge  *v.* 使突然地下落；猛插；骤降；陡峭地向下倾斜；颠簸；跳进；（使）陷入；栽种
teleconference  电话会议；远程会议
theoretical  *adj.* 理论的；理论上的；假设的；推理的
weaken  *vt.* 减少；使变弱；使变淡
arguably  *adv.* 可论证地；可争辩地；正如可提出证据加以证明的那样地；可能，大概
correlated  *adj.* 有相互关系的
forefront  *n.* 最前线，最前部；活动的中心；（思考、关注的）重心

production efficiency　生产效率
negotiation　*n.* 谈判；转让；顺利的通过
falter　*vi.* 衰弱；衰退；(嗓音)颤抖；结巴地说；蹒跚；犹豫
feedback　*n.* 反馈；成果，资料；回复
clear up　清理；放晴；整理；打扫
frequent　*adj.* 频繁的；时常发生的；惯常的
observer　*n.* 观察者；观测者；遵守者
commentator　*n.* 评论员，解说员；实况播音员；时事评论者
fortune　*n.* 财富；命运；运气
outperform　*vt.* 胜过；做得比……好
facilitate　*vt.* 促进；帮助；使容易
modernized　*adj.* 现代化的
Arabian Nights　一千零一夜(书名，又名天方夜谭)；不真实的故事
managerial　*adj.* 管理的；经理的
constituent　*adj.* 构成的；选举的；有任命(或选举)权的；立宪的；有宪法制定(或修改)权的
implement　*vt.* 实施，执行；实现，使生效
foreseeable　*adj.* 可预知的，可预测的
palpably　*adv.* 易觉察地；可触知地
rival　*n.* 竞争对手，敌手，竞争者
characteristically　*adv.* 典型地；表示特性地
privacy-friendly　保护隐私的或有利于隐私保护的
attach　*vi.* 附加；附属；伴随
intangible　*adj.* 无形的，触摸不到的；难以理解的
transit　*vt.* 运送
contentment　*n.* 满足；满意
productivity　*n.* 生产力；生产率；生产能力
magnifier　*n.* 放大镜；放大器
turbojet　*n.* 涡轮喷气飞机
favorable　*adj.* 有利的；良好的；赞成的，赞许的；讨人喜欢的
unparalleled　*adj.* 无比的；无双的；空前未有的
profitability　*n.* 盈利能力；收益性；利益率
earnings per share　每股收益，每股盈余，每股盈利
labor productivity　劳动生产率
beneficial　*adj.* 有益的，有利的；可享利益的
prone to　有……倾向的
shift　*v.* 转移；快速移动；变换；改变观点；推卸(责任)；振作；移位；狼吞虎咽地吃；去除(污迹)；销售，出售；换挡；轮班；含糊其词，拐弯抹角
transform　*vt.* 改变，使……变形；转换
intercontinental　*adj.* 洲际的；大陆间的

immune　*adj.* 免疫的；免于……的，免除的
depressed　*adj.* 沮丧的；萧条的；压低的
up and down　上上下下；到处；前前后后；来来往往

## Notes

1. The regular transport means that the transportation is operated in accordance with the confirmed timetable and its time and itinerary is expressly indicated without any need for particular explanation, such as airway transport, railway transport, highway transport and waterway transport, etc.

— in accordance with　依照；与……一致

译文：定期运输方式就是指依照已确认的时刻表进行运营，不需要任何特别说明就能非常明确地表示时间和行程，如航线运输、铁路运输、公路运输和水路运输等。

2. On-demand air transport brings advantages and benefits to its variety of users including individuals, company managers, troubleshooters, sport teams, entrepreneurs and some families. They share one same goal that they could reach a certain destination as safely and comfortably as possible.

— brings benefits to　带来效益

— troubleshooter　*n.* 解决纠纷者；故障检修工；（计算机）故障查找软件

译文：有各种人员在使用按需航空运输：个体人员、公司管理人员、技术排故人员、体育运动队、企业家及一些家庭，他们全都从这种飞行运输模式的使用中受益。他们的共同目标是：尽可能安全和舒适地到达某个旅行目的地。

3. Some take it a sign of their social status; some take more account of its economical benefit; some prefer a faster or recreational travel; some thrill-seekers challenge to pilot themselves, while others would prefer to outsource the flight missions to professional flight crew.

— social status　社会地位，身价

— take account of　考虑到；顾及；体谅

— thrill-seeker　寻找惊险刺激的人

— flight crew　机组人员；飞行人员；空勤人员

译文：有人以此显示自己的身份，有人更注重经济实惠；有人为了快捷，有人更注重休闲消遣；有人为追求刺激而亲自驾驶飞机旅行，有人则将飞行任务交给专业的飞行机组来执行。

4. Business aviation encompasses all types of air flights for the purpose of business activities, under which the main types are employee-flown operations and corporate operations.

译文：公务航空包括以公务活动为目的的所有类别的航空飞行，在这个大的分类之下，主要的类型是员工自驾式运营(employee-flown operations)和企业式运营 (corporate operations)。

5. NBAA (National Business Aviation Association) defines corporate aviation as follows: to promote business, a company or enterprise undertakes to operate self-

owned or leased aircrafts to carry staff or commodities by employing professional pilots who will directly receive salary or special rewards on flight task basis.

— NBAA  *abbr*. 美国国家公务航空协会（National Business Aviation Association）。美国国家公务航空协会（NBAA）创立于1947年，总部设于华盛顿特区，它为依靠民用飞机提高效率和生产率并获取成功的公司提供服务，并且是其中翘楚。

译文：美国国家公务航空协会（NBAA）是这样定义"企业航空"的："自有的或租赁的飞机，由一个公司或者企业组织运行，为促进其经营来运送职员或货物，雇用专业飞行员，并且这些飞行员以飞行任务直接取得薪水或特殊报酬"。

6. While business intercouse crowded with phone calls, emails, faxes and postal mails, people are also in need of direct in-person communications to finalize business contracts, address thorny issues and reach new consensus.

— business intercourse  商业往来
— in-person  亲自，在场的，亲身的
— reach a consensus  达成共识，达成一致意见

译文：在电话、电子邮件、传真及邮政信件用于商业往来的同时，人们还需要在面对面的直接交流中完成贸易合同，解决棘手问题，达成新的共识。

7. According to the prediction of technical experts, business travel will plunge due to the emergence of mobile phones, networks and teleconferencing. Their theoretical research points out that better and newer communication equipment weakens direct in-person talks and teleconference will rise as a best alternative.

— theoretical research  理论研究，理论探讨

译文：技术专家曾预测，公务旅行将会因为手机、网络和电话会议的出现而锐减。他们的理论研究指出：更好、更新的通信设备使得面对面的直接会谈已不再重要，而电话会议将是一个最好的方式。

8. Through meetings, we can build up friendship with customers, understand their needs and feedbacks, help them clear up problems, and introduce new products and services. Business travel is a long-term industry.

— build up friendship  建立友谊
— clear up  清理；放晴；整理；打扫

译文：通过会晤，我们可以与客户建立友谊，了解客户的需求，听取反馈意见，并且帮助他们解决问题，还可以引进新产品和服务。公务旅行是一个长期发展的产业。

9. In 2002, more than 13000 operators deployed a total of above 21000 gas turbine (turbojet and turboprop) aircrafts to the business aircraft market. And three quarters of these operators and aircrafts were based in the North America.

— turbojet  *n*. 涡轮喷气发动机；涡轮喷气飞机。特点是完全依赖燃气流产生推力。通常用作高速飞机的动力，但油耗比涡轮风扇发动机高。

译文：2002年，世界范围内13000多家运营商投放了21000多架燃气涡轮（涡喷和涡桨）式飞机加入公务机运营，3/4的运营商和飞机集中在北美。

10. Of the Fortune 500 companies, only 29% don't have their own business aircrafts. And the companies in possession of business aircrafts greatly outperform

those without the aircrafts in terms of their productivity, sales revenue and profits. The companies take business aircraft as a tool to improve productivity, through which they can take control of the tight travel schedule and facilitate the daily traffic of their business managers who need to attend every production activity.

—《财富》(Fortune Magazine) 是一本由美国人亨利·鲁斯创办于1930年,主要刊登经济问题研究文章的杂志。现隶属时代华纳集团旗下的时代公司。《财富》杂志自1954年推出全球500强排行榜,历来都成为经济界关注的焦点,影响巨大。《财富》杂志举办了一系列引人注目的财经论坛,如著名的《财富》全球论坛,即世界500强年会便是其中之一。

译文:在《财富》500强企业中,只有29%的企业没有自己的公务机。拥有公务机的企业其生产率、销售收入和利润大大超过那些没有公务机的企业。这些企业将公务机作为一种提高生产率的工具,通过它来控制繁忙的旅行时间表,并为那些需要考虑每项生产活动的企业管理者们每天的工作往来提供交通便利。

11. In comparison, those enterprises owning business aircrafts are palpably in better position than their rivals, because air transport by business aircrafts makes business operation characteristically time-saving, flexible, highly efficient, privacy-friendly and safe, and in the same time naturally attaches intangible benefits, such as longer relaxing time at home for travelers, and more comfortable non-stop or minimum-transit travel to the farthest destination.

— time-saving　节省时间
— privacy-friendly　保护隐私
— intangible benefits　无形的好处

译文:自己拥有飞机的企业与没有公务机的竞争对手相比,所具有的优势是显而易见的。企业航空运输的省时、灵活、高效、保护隐私和安全等特性,足以体现公务机在企业运营中的特殊作用。而且,更加舒适和无形的好处自然也会附加到它的应用上。例如,增加旅行者在家的休息时间,有足够能力直接到达最遥远的地方,舒适而且中途转机次数最少。

## Exercises

### Section one: Answer the following questions according to the text.

1. In comparison with civil airlines, what are the advantages or benefits of business aircraft?
2. What do you think of the status of business aircraft in United States?
3. What are the main types of engines used on business aircrafts?
4. Why do we still need another aircrafts to transport a small number of passengers to the same destination? Isn't it expansive? And how safe is it?
5. What impacts does emergence of mobile phones, networks and teleconferencing bring to business travel?

Section two: Match the words or phrases in column A with the definition in column B.

A
1. seize
2. enterprise
3. commodity
4. timetable
5. outsource
6. waterway
7. highway
8. frequent
9. productivity
10. competitive

B
a. a schedule of times of arrivals and departures
b. to contract work out
c. a navigable body of water
d. take hold of; grab
e. an organization created for business ventures
f. involving competition or competitiveness
g. coming at short intervals or habitually
h. the quality of being productive or having the power to produce
i. a major road for any form of motor transport
j. articles of commerce

Section three: Translate the following into Chinese.

1. Research shows that companies that own airplanes, whether in terms of sales volume, earnings per share, long-term returns to investors, or labor productivity (sales per employee), are growing faster than those that don't have corporate jets.

2. All of these are calling for the more important role of business jet operation, or in other words, the status quo of the airlines may be the best reason for travelers to shift to business jets.

3. Global village is gradually transformed to an international market in which the application of business aircraft will greatly facilitate such transition.

4. When the business is booming, there will be more flight activities; when the business is depressed, not only will there be less flight, but also there will be less flight time and personnel.

5. The flight will present ups and downs with the trend of the financial situation, but the operation of business aircraft services will sustain and grow even more powerful with the deeper integration of on-demand aviation operation into the corporate culture.

6. In fact, business aircraft turns out to be as much as the modernized magic time machine in the Arabian Nights, bringing managerial personnel safely and quickly to the next business opportunity.

7. In 2002, more than 13000 operators deployed a total of above 21000 gas turbine (turbojet and turboprop) aircrafts to the business aircraft market.

Section four: Translate the following into English.
1. 她尽力显得友好，可是显然很拘束。
2. 国内舆论已转而反对战争。

3. 这些文件已向公众开放。
4. 这家公司的商业前景看起来非常光明。
5. 另一个频道对奥运比赛的实况报道要好得多。
6. 报纸上有很多艺术方面的报道,可是政治时事述评却不多。
7. 数位消防人员因表现英勇而受到了表彰。
8. 他的女儿因成就杰出获得了嘉奖,他为此感到无比自豪。
9. 我们将于明年8月破土动工。
10. 作为一部低预算影片,它有不少值得称道之处。

# 3.4 Small aircraft
# 小型飞机

## Background

### Introduction to small aircraft

Small aircrafts with 2 to 6 seats are the most common and one of the most active kinds in GA aircraft market. Based mainly in North America, Western Europe and other developed places, they are primarily manufactured by small independent companies which usually focus on making only a few ranges of aircraft models and their derived models. These aircrafts, cheap in price, are mostly owned by individuals and a small part of them are owned for special missions. They are commonly used for private aircraft, training, sightseeing, administration, aviation sports, aerobatics, forest patrol and law enforcement. And the annual Air Adventure general aircraft exhibition held by the American Experimental Aircraft Association (EAA) in Oshkosh, Wisconsin is known as a grand event of general aircraft in this kind.

Aircrafts in this kind are usually configured with one piston engine, whereas a few are engined with double pistons (such as the Diamond's DA-42 Twin Star) or are replaced with one turboprop engine instead, though the overall designs are not changed (such as the Piper Aircraft's Malibu Meridian updated from the PA-46-350P Malibu Mirage by replacing one PT6A engine). More recently, the light general-purpose diesel-engined aircrafts also turn up in the market, like the Skyhawk equipped with Thielert's 120kW Centurion 2.0 diesel engine.

The development of lightweight aircraft had fallen into decline for a time, but soon rebounded in American market as with the signing of General Aviation Revitalization

Act by the Clinton Administration on August 12, 1994, the interests of general aviation and other industries was legislatively adjusted.

## Major technical features

Small aircrafts in above kind mainly adopts conventional layout. Except for the dual-engine low-wing arrangement of DA-42 Twin Star under the Diamond Aircraft, small aircrafts are basically equipped with one piston engine. The main piston aeroengine manufacturers in Europe and the United States all specialize in the production of horizontally opposed engines, with the few in-line engines and radical engines available only in Eastern European countries. And there're mainly two types of horizontally opposed engines, namely of 4-cylinder and of 6-cylinder, so as to meet the requirements of different aircrafts. Also, given that the small aircrafts manufacturers are mainly USA-based, therefore the main piston engine companies include Teledyne Continental Motors (with engine's power range between 74.6–275.9kW) and Textron Lycoming (with engine's power range between 85.8–317kW) only.

The small single-engine aircrafts are laid out mainly with either high wing or low wing. Most of high-winged aircrafts are equipped with stay rods and their landing gears cannot be retracted, and in some models the landing gears have no fairing or are just pontoons. As to the low-winged aircrafts, their landing gears can be retractable and non-retractable subject to the flight speed and wing's thickness, and those with non-retractable gears are by and large configured with fairings. In general, the low-winged aircrafts outperform the high-winged aircrafts. And the majority of small aircrafts are made with cruciform tail (or T tail) and nose-wheel landing gears, though some may adopt the tail-wheel landing gears.

As Cessna builds a competitive edge in the market of high-winged aircrafts, seldom are there any other models of high-winged aircrafts except for a handful of simple versions in that kind. Once put on the market at a price of USD 109,500, Cessna's 162 Skycatcher, with its advantages in both output and price, almost squeezes out of every possible rivals. Hence Cessna's leading rivals such as Cirrus Design and Columbia Aircraft all turn to the market of low-wing layout and work to achieve a better flight performance through constructive aerodynamic layout. In addition, Diamond Aircraft also manages to open up a differentiated market by making a model that is between the ultra light aircraft and the small aircraft. Currently the Columbia Aircraft is filing for bankruptcy due to financial distress, and Cessna, taking the occasion, is trying to acquire Columbia Aircraft so as to make inroads into the market of advanced small low-winged aircrafts. However, as Cirrus Design also steps in the case, the market will be on wait-and-see. As to other market players Piper Aircraft and Hawker Beechcraft, they, with the similar operation of administration aircraft, business aircraft and sightseeing aircraft, are holding other segmented markets through better performance, higher comfort and longer flying range.

### Applications of small aircrafts

Small aircrafts are used in pilot training, aviation sports, aerial tourism, education and training, culture and sports, aviation advertising, etc.

Pilot training: in North America, for example, there were 600,000 people with flight licenses in the United States in 2005. With a multitude of flying cadets, it is obviously uneconomical to train them with large aircrafts; instead, small aircrafts should be counted on. Furthermore, aircraft manufacturers and some dealers also establish pilot training as one of the main approaches to expand sales. For instance, Cessna had instituted a worldwide flight training network including 225 pilot centers, and in the more than 20 years from 1977 to 2006, about 400,000 pilots were trained, which in turn greatly promoted the sales of aircrafts.

Generally speaking, training cost differs, depending on the selection of the model of trainer aircraft, the qualification grade of trainers as well as the training courses. Examples are numerous. Training with pontoon-equipped Cessna 172 or 180 is charged at around USD 1,100 per hour while training with standard Cessna 172 aircraft costs about USD 780 per hour. In North America, you don't have to pay through the nose to obtain a commercial flight license. The international airline training company in Canada only charges 8,770 Canadian dollars, or an equal amount of 47,000 yuan, provided that the trainee obtains private flight license first before starting the commercial flight training, which in total costs less than 100,000 yuan.

Leisure and sports: in private aviation, globally there're hundreds of thousands of small airplanes used for leisure and recreation, and meanwhile hundreds of thousands of flight fans are also built up, creating a huge niche market.

Pilot certificate: Federal Aviation Administration (FAA) in the United States issues different categories of certificate to new pilots according to the results of their paper test and flight test, which are all conducted under the supervision of FAA representatives. At the moment, there are totally four categories of pilot certificate, the latest one for recreational pilot and the other three respectively for private pilot, commercial pilot and airline transport pilot. It is worth noting that all pilots must fly first solo at a certain time, and all future pilots must first obtain a private pilot certificate and then an instrument rating certificate. As for certificates of other categories, trainees may apply accordingly based on their personal lifestyle or job requirements.

### Typical models

1. LE500

### Profile

LE500 Aircraft, a brain child of China's own intellectual property rights, is a light multi-purpose aircraft jointly developed and produced by Shijiazhuang Aircraft

Industry Co., Ltd. of AVIC Ⅱ, The First Aircraft Institute of AVIC Ⅰ and Civil Aviation Flight University of China (CAFUC). The Aircraft made successfully its maiden flight on October 26, 2003, a sign that China's gap in the manufacturing of 4-or-5-seater light multi-purpose GA aircraft was filled.

LE500 Aircraft is China's first commercial aircraft for private use. It is a new type of GA aircraft designed by advanced digital means and thoughtfully developed in strict accordance with CAAC's CCAR-23 airworthiness regulations. The comprehensive performance of the aircraft equals the advanced level of aircraft in the same model around the world, and it has an apparent competitive advantage in the market of general-purpose light aircraft as it excels other currently imported similar aircrafts in terms of flight performance, ride comfort, interior decoration, after-sale service and economical efficiency.

## 2. Cessna 162 Skycatcher

### Profile

On July 22, 2007, Cessna Aircraft said in a press release at the Oshkosh air show that the Cessna light two-seater sport aircraft, after over a year of planning, was launched and open to pre-order. The aircraft, modeled 162, is named as Skycatcher. The 162 Aircraft features a brand-new design and is a full-metal propeller-induced sports aircraft configured with piston engine, high-wing layout, fixed landing gear, and side-by-side two seats (including pilot seat). It can be used for aerobatic performance, rudimentary training, flight club or private ends.

Thanks to its low purchasing and operational costs, the aircraft takes great hold of a promising market. By pricing policy, the order price of the first 1000 standard equipped 162 Aircraft is USD 109,500 per aircraft, and the down payment is USD 10,000. And the price after the $1000^{th}$ aircraft will be increased to USD 115,000 each. Two days after the news release, Cessna announced on July 24 that it had received more than 400 orders, and the delivery was planned to start in 2009, with an annual production of 700 aircrafts.

### 3. Cessna 172 Skyhawk and 182 Skylane

### Profile

These aircrafts are classics of Cessna Aircraft. They are single engine propeller-induced fixed high-wing metallic aircrafts (4-seater). Among them, more than 43,000 172-Skyhawk have been delivered.

As a typical product, the single-piston-engined light propeller aircraft has been used to train pilots from generation to generation, and many of which produced two decades ago are now still in good service. Over the decade from 1986 to 1995, Cessna suspended the production of light piston aircraft. However, under the call of new regulations, Cessna in 1996 reconstructed the production line of the single-piston-engine aircraft and resumed the production after improvement of design in accordance with FAR-23 standard. And a year later, Cessna started supplying a new generation of Skyhawk aircraft. The 4-seater 172R, 172S, 182T and T182T aircrafts mainly target at flight school and individual aviation fans. For 172 and 182 aircraft, their main difference lies in the type of engine and propeller assembled in which the 172 aircraft is two-bladed and the 182 aircraft is three-bladed.

Honorably titled as the world's most popular model, this aircraft enjoys an excellent safety performance and has less chance of occurring serious flight accident when opposed to other aircrafts in same model. Generally speaking, this aircraft is more often used as trainer aircraft, private aircraft and business aircraft.

In the mid-2005, CAFUC entered into a contract with Cessna Aircraft for the procurement of 42 single-engine-piston light 172R-Skyhawk aircrafts. And in the year of 2004, Cessna's standard 172 Skyhawk and 182 Skylane cost respectively USD 158,000 and USD 250,000.

Unit 3　GA aircrafts and manufacture 通用航空飞机及制造

## General Vocabulary

derived　*adj.* 导出的；衍生的，派生的
aerobatics　*n.* 特技飞行；特技飞行术
adventure　*n.* 冒险；冒险精神；投机活动
kW　*abbr.* 千瓦特（kilowatt、kilowatts）
EAA(American Experimental Aircraft Association)　美国实验飞机协会
grand　*adj.* 宏伟的；豪华的；极重要的
replace　*vt.* 取代，代替；替换，更换；归还，偿还；把……放回原处
turboprop　*n.* 涡轮螺桨发动机；涡轮螺旋桨飞机
diesel　*n.* 柴油机；柴油；(俚) 健康的身体
rebound　*v.* (球或其他运动物体) 弹回，反弹；(价格、价值等下跌后) 回升，反弹；抢 (篮板球)；(事件，局势) 产生事与愿违的结果
legislatively　*adv.* 立法地
adjust　*vt.* 调整，使……适合；校准
adopt　*vt.* 采取；接受；收养；正式通过
aeroengine　*n.* 航空发动机；飞机引擎

specialize　*vi.* 专门从事；详细说明；特化
horizontally　*adv.* 水平地；地平地
in-line　*adj.* 一列的，直排的；同轴的；（计算机程序）嵌入的，内嵌的；构成完整连续作业一部分的；（生产过程）顺序连接的
radical　*adj.* 激进的；根本的；彻底的
cylinder　*n.* 圆筒；汽缸；柱面；圆柱状物
high wing　上单翼
low wing　下单翼
stay rod　牵条螺栓，缀条，撑杆，锁定杆
retract　*vt.* 缩回；缩进；取消
fairing　*n.* 整流罩；集市上卖的礼物；酬谢礼品
pontoon　*n.* 浮筒；浮码头；浮舟；驳船
thickness　*n.* 厚度；层；浓度；含混不清
by and large　大体上，总的来说
outperform　*vt.* 胜过；做得比……好
cruciform　*adj.* 十字形的；十字架状的
nose-wheel　前轮；机头前轮；前舱
tail-wheel　尾轮
seldom　*adv.* 很少，不常
handful　*n.* 少数；一把；棘手事
squeeze　*v.* 挤；紧握；勒索；压榨；使挤进；（非正式）向……施加压力；（由于金融或商业）破坏
rival　*n.* 竞争对手；可与……匹敌的人；同行者
constructive　*adj.* 建设性的；推定的；构造上的；有助益的
aerodynamic　*adj.* 空气动力学的，航空动力学的
differentiated　*adj.* 分化型；已分化的；可区分的
ultra light aircraft　超轻型飞机
bankruptcy　*n.* 破产
distress　*n.* 危难，不幸；贫困；悲痛
occasion　*n.* 时机，机会；场合；理由
wait-and-see　*adj.* 观望的
segmented　*adj.* 分段的
flying cadet　飞行学员
uneconomical　*adj.* 不经济的；浪费的；不节俭的
dealer　*n.* 经销商；商人
approach　*n.* 方法，方式；接近；接洽；（某事的）临近；路径；进场（着陆）；相似的事物
qualification　*n.* 资格；条件；限制；赋予资格
numerous　*adj.* 许多的，很多的
pay through the nose　花很多钱

trainee　*n.* 练习生，实习生；受训者；新兵；训练中的动物
leisure　*n.* 闲暇；空闲；安逸
niche market　瞄准机会的市场；缝隙市场
certificate　*n.* 证书；文凭，合格证书；电影放映许可证
Federal Aviation Administration (FAA)　联邦航空管理局
category　*n.* 种类，分类；范畴
representative　*n.* 代表；典型；众议员；销售代表
solo　*adv.* 单独地
accordingly　*adv.* 因此，于是；相应地；照着
lifestyle　*n.* 生活方式
brain child　*n.* 脑力劳动的产物
intellectual　*adj.* 智力的；聪明的；理智的
AVIC　*abbr.* 中国航空工业集团公司（Aviation Industry of China）
Civil Aviation Flight University of China (CAFUC)　中国民航飞行学院
maiden　*adj.*（尤指年纪较大女性）未婚的；处女的；（航行、飞行）首次的；（赛马）从未跑赢过的；（植物）生长期第一年内的
digital　*adj.* 数字的；手指的
thoughtfully　*adv.* 沉思地；体贴地，亲切地
CAAC　*abbr.* 中国民用航空总局（General Administration of Civil Aviation of China）
CCAR　*abbr.* 中国民航规章（China Civil Aviation Regulations）
airworthiness　*n.* 适航性；耐飞性
apparent　*adj.* 显然的；表面上的
competitive advantage　竞争优势
excel　*vt.* 超过；擅长
interior　*adj.* 内部的，里面的；内位的；内陆的，腹地的；内务的，内政的；心灵的，精神的；本质的
decoration　*n.* 装饰，装潢；装饰品；奖章
after-sale　售后服务，售后的
press release　新闻稿；通讯稿
pre-order　*vt.* 预购；预订
full-metal　全金属
side-by-side　*adj.* 并肩的；并行的
rudimentary　*adj.* 基本的；初步的；退化的；残遗的；未发展的
flight club　飞行俱乐部
down payment　（分期付款中的）头期款；预付定金
delivery　*n.* 交付；分娩；递送
classic　*n.* 名著；经典著作；大艺术家
metallic　*adj.* 金属的，含金属的
suspend　*vt.* 延缓，推迟；使暂停；使悬浮
assemble　*vt.* 集合，聚集；装配；收集

blade　　*n.* 叶片；刀片，刀锋；剑
honorably　　*adv.* 体面地；值得尊敬地
procurement　　*n.* 采购；获得，取得

## Notes

1. They are commonly used for private aircraft, training, sightseeing, administration, aviation sports, aerobatics, forest patrol and law enforcement. And the annual Air Adventure general aircraft exhibition held by the American Experimental Aircraft Association (EAA) in Oshkosh, Wisconsin is known as a grand event of general aircraft in this kind.

　　– private aircraft　私人飞机

　　– American Experimental Aircraft Association (EAA)　美国实验飞机协会 EAA 是美国航空爱好者自发的群众性的飞行大会，全称是全美实验飞机协会。每年的夏季都隆重举行一次，每次都有上万架大小各异的飞机和近百万狂热的美国航空爱好者参加这个飞行盛会，其中绝大多数是私人拥有的或是自制的小型飞机。EAA 展现了美国民众强烈的航空意识和航空热情，是美国群众性的航空活动发展到一定规模后的必然产物，也显示了美国这个航空大国背后的群众基础。EAA 也能从一个侧面说明美国之所以能够成为世界一流的航空大国的原因所在。

　　译文：常见的用途有私人飞机、教练、游览/观光/行政、航空运动、特技飞行、森林巡逻与执法等。此外，不少个人也自己制造这一级别的飞机。美国实验飞机协会（EAA）在威斯康星州奥什科什城举办的一年一度的"空中冒险"通用飞机展是这一级别通用飞机的盛会。

2. The development of lightweight aircraft had fallen into decline for a time, but soon rebounded in American market as with the signing of General Aviation Revitalization Act by the Clinton Administration on August $12^{th}$, 1994, the interests of general aviation and other industries was legislatively adjusted.

　　– General Aviation Revitalization Act　《通用航空振兴法案》
　　– the Clinton Administration　克林顿政府

　　译文：轻型飞机的发展市场一度衰退，但是克林顿政府 1994 年 8 月 12 日签署了《通用航空振兴法案》，用立法手段调整通用航空与其他行业的利益关系，迅速扭转了当时美国通用航空下滑的趋势。

3. Except for the dual-engine low-wing arrangement of DA-42 Twin Star under the Diamond Aircraft, small aircrafts are basically equipped with one piston engine.

　　– dual-engine　双发，两个发动机
　　– low-wing　下单翼　下单翼飞机是指机翼安装在机身下部的飞机，下单翼飞机的优点是机翼翼梁穿过机舱，机翼强度高，阻力小，升力大。主起落架布置在机翼根部，强度较高，起落架舱又可以设置在机翼根部的整流罩中，翼吊发动机距离地面较近，从而方便维护，飞机机翼还可以用来作为紧急撤离时的通道，高效地撤离旅客。

　　译文：除去钻石公司的 DA-42"双子星"采用双发下单翼布局外，小型飞机基本上都装一台活塞发动机。

4. And there're mainly two types of horizontally opposed engines, namely of

4-cylinder and of 6-cylinder, so as to meet the requirements of different aircrafts.

- horizontally opposed engines　水平对置发动机，发动机活塞平均分布在曲轴两侧，在水平方向上左右运动。使发动机的整体高度降低、长度缩短、重心降低，使行驶更加平稳，发动机安装在整体的中心线上，两侧活塞产生的力矩相互抵消，大大降低飞机在行驶中的振动，使发动机转速得到很大提升，减少噪声。

译文：为满足不同级别飞机的需要，水平对置发动机主要有4缸、6缸两种形式。

5. The small single-engine aircrafts are laid out mainly with either high wing or low wing. Most of high-winged aircrafts are equipped with stay rods and their landing gears cannot be retracted, and in some models the landing gears have no fairing or are just pontoons.

- high wing　上单翼。上单翼飞机是指机翼安装在机身上部的飞机，上单翼飞机的最大优点是机场适应性好。这种飞机的机翼离地高度高，机翼下面有足够的空间吊挂发动机，不会轻易地将地面的沙石吸入进气道，损坏发动机。这对军用运输机、轰炸机而言尤为重要。军用机场的条件千差万别，跑道上难免会有尘土、泥沙和碎石，所以上单翼布局对防止发动机吸入异物有较好的作用。由于没有发动机离地间隙的限制，可以将机身设计得离地很近，大幅度降低货仓地板的高度，适合装运大件货物。

- fairing　整流罩。飞机上罩于外突物或结构外形不连续处以减少空气阻力的流线型构件。飞机的整流罩顾名思义就是用于调整气流的，为了减少在飞机在空中的阻力，整流罩尽量以流线型设计，使气流能平滑的通过。此外，如果没有整流罩，发动机的管路、机匣和核心机都会暴露在外面，从而可能使发动机被外来物撞击造成损伤，而有了整流罩就能很好地保护发动机内部组件。

译文：小型飞机单发机型布局主要有以下两种：上单翼或下单翼布局。其中，上单翼机型多数带有撑杆，起落架不可收放，部分机型起落架没有整流罩，或起落装置为浮筒。

6. In general, the low-winged aircrafts outperform the high-winged aircrafts. And the majority of small aircrafts are made with cruciform tail (or T tail) and nose-wheel landing gears, though some may adopt the tail-wheel landing gears.

- tail-wheel landing gear　后三点式起落架，后三点式起落架的两个支点(主轮)对称地安置在飞机重心前面，第三个支点(尾轮)位于飞机尾部。

- nose-wheel landing gear　前三点式起落架，前三点式起落架的两个支点(主轮)对称地安置在飞机重心后面，第三个支点(前轮)位于机身前部，尾部通常还装有保护座，防止在飞机离地时出现擦尾。

译文：但整体上，下单翼机型性能优于上单翼机型。通常大多数小型飞机为前三点式起落架，也可采用后三点式起落架。尾翼为传统的十字形尾翼或T形尾翼。

7. Once put on the market at a price of USD 109,500, Cessna's 162 Skycatcher, with its advantages in both output and price, almost squeezes out of every possible rivals. Hence Cessna's leading rivals such as Cirrus Design and Columbia Aircraft all turn to the market of low-winged layout and work to achieve a better flight performance through constructive aerodynamic layout.

- squeeze out　挤出；榨出；排出

- aerodynamic layout　气动布局，气动布局同飞机外形构造、大部件的布局、飞机的

127

动态特性及所受到的空气动力密切相关。关系到飞机的飞行特征及性能。故将飞机外部总体形态布局与位置安排称作气动布局。简单地说，气动布局就是指飞机的各翼面，如主翼、尾翼等是如何放置的，气动布局主要决定飞机的机动性，至于发动机、座舱以及武器等放在哪里的问题，则笼统地称为飞机的总体布局。

译文：而鉴于赛斯纳飞机公司的产量和价格优势，在162"天捕手"（Skycatcher）以10.95万美元的价格投入市场后，这一市场已基本没有竞争对手生存的空间。主要竞争对手均采取下单翼布局，并通过良好的气动布局以获得更好的飞行性能。

8. In North America, you don't have to pay through the nose to obtain a commercial flight license. The international airline training company in Canada only charges 8,770 Canadian dollars, or an equal amount of 47,000 yuan, provided that the trainee obtains private flight license first before starting the commercial flight training, which in total costs less than 100,000 yuan.

— pay through the nose　　花很多钱
— private flight license　　私人飞行执照，持有人可以驾驶其私人航空器进行非盈利的飞行活动，取得执照所需的条件由国际民航组织（ICAO）规定。根据ICAO规定，申请者必须顺利完成在空中至少40小时（英国和西班牙规定为45小时）的飞行培训，通过笔试，完成长时间的单人飞行（最低时间要求为10小时）等。不过在实际的执行过程中，各国间的要求有很大不同。

译文：在北美，要拿到商业飞行执照也不是很贵，如加拿大的航线培训国际公司，总共只收8770加元，约合人民币47000元。当然，学员应首先取得私人飞行执照，才能开始商业飞行培训。即便把这两项加在一起，也不超过10万元人民币。

9. Leisure and sports: in private aviation, globally there're hundreds of thousands of small airplanes used for leisure and recreation, and meanwhile hundreds of thousands of flight fans are also built up, creating a huge niche market.

— niche market　　瞄准机会的市场；缝隙市场

译文：休闲运动：私人航空，全世界用于休闲娱乐的小型飞机数以十万计，培养了数十万名飞行爱好者，并创造了巨大的市场空间。

10. Pilot certificate: Federal Aviation Administration (FAA) in the United States issues different categories of certificate to new pilots according to the results of their paper test and flight test, which are all conducted under the supervision of FAA representatives.

— paper test　　笔试

译文：飞行员合格证书：联邦航空管理局（FAA）颁发给新飞行员的合格证书分为不同的类别，这要根据飞行员笔试和飞行测试结果而定，笔试和飞行测试都是在FAA代表的监督下进行。

11. LE500 Aircraft, a brain child of China's own intellectual property rights, is a light multi-purpose aircraft jointly developed and produced by Shijiazhuang Aircraft Industry Co., Ltd. of AVIC Ⅱ, The First Aircraft Institute of AVIC Ⅰ and Civil Aviation Flight University of China (CAFUC).

— brain child　　脑力劳动的产物

— intellectual property right  知识产权；智慧财产权

— AVIC  中国航空工业集团公司（Aviation Industry of China）的缩写，是中国国家出资设立，由国务院国资委代表国务院履行出资人职责的国有独资公司，是由中央管理的国有特大型企业，也是国家授权投资的机构，于2008年11月6日由原中国航空工业第一集团公司和中国航空工业第二集团公司重组整合成立。中国航空工业集团有限公司设有航空武器装备、军用运输类飞机、直升机、机载系统与汽车零部件、通用航空、航空研究、飞行试验、航空供应链与军贸、资产管理、金融、工程建设、汽车等产业。

译文："小鹰"500飞机是中国航空工业第二集团公司石家庄飞机工业有限责任公司、中国航空工业第一集团公司第一飞机设计研究院、中国民航飞行学院联合研制生产的，具有我国自主知识产权的轻型多用途飞机。

12. The comprehensive performance of the aircraft equals the advanced level of aircraft in the same model around the world, and it has an apparent competitive advantage in the market of general-purpose light aircraft as it excels other currently imported similar aircrafts in terms of flight performance, ride comfort, interior decoration, after-sale service and economical efficiency.

— interior decoration  内饰
— after-sale service  售后服务

译文：该机综合性能达到世界同类机型先进水平，主要飞行性能、驾乘舒适度、内装饰造型、售后服务及使用经济性，均优于目前进口的国外同类飞机，在我国轻型通用飞机的市场竞争中具有明显优势。

# Exercises

Section one: Answer the following questions according to the text.

1. Why are small aircrafts most common in GA aircraft market, especially in North America?
2. How did the development of lightweight aircraft rebound after a period of recession?
3. What are the main technical features of small aircraft?
4. List some major applications of small aircrafts.
5. Please briefly demonstrate some typical models of small aircrafts.

Section two: Match the words or phrases in column A with the definition in column B.

| A | B |
| --- | --- |
| 1. application | a. a movement back from an impact |
| 2. maiden | b. a state of adversity (danger or affliction or need) |
| 3. comprehensive | c. a plan or design of something that is laid out |
| 4. adjust | d. the act of bringing something to bear |
| 5. deliver | e. following accepted customs and proprieties |

6. certificate      f. serving to set in motion
7. distress      g. including all or everything
8. conventional      h. bring to a destination, make a delivery
9. layout      i. alter or regulate so as to achieve accuracy or conform to a standard
10. rebound      j. a document attesting to the truth of certain stated facts

### Section three: Translate the following into Chinese.

1. Based mainly in North America, Western Europe and other developed places, they are primarily manufactured by small independent companies which usually focus on making only a few ranges of aircraft models and their derived models.

2. The development of lightweight aircraft had fallen into decline for a time, but soon rebounded in American market as with the signing of General Aviation Revitalization Act by the Clinton Administration on August 12, 1994, the interests of general aviation and other industries was legislatively adjusted.

3. As to the low-winged aircrafts, their landing gears can be retractable and non-retractable subject to the flight speed and wing's thickness, and those with non-retractable gears are by and large configured with fairings.

4. Generally speaking, training cost differs, depending on the selection of the model of trainer aircraft, the qualification grade of trainers as well as the training courses.

5. At the moment, there are totally four categories of pilot certificate, the latest one for recreational pilot and the other three respectively for private pilot, commercial pilot and airline transport pilot.

6. On July 22, 2007, Cessna Aircraft said in a press release at the Oshkosh air show that the Cessna light two-seater sport aircraft, after over a year of planning, was launched and open to pre-order.

7. Thanks to its low purchasing and operational costs, the aircraft takes great hold of a promising market. By pricing policy, the order price of the first 1000 standard equipped 162 Aircraft is USD 109,500 per aircraft, and the down payment is USD 10,000.

8. As a typical product, the single-piston-engined light propeller aircraft has been used to train pilots from generation to generation, and many of which produced two decades ago are now still in good service.

### Section four: Translate the following into English.

1. 为纪念诗人百年诞辰建起了一尊雕像。
2. 她站在领奖台上时泪水顺着她的脸庞流了下来。
3. 你应该用一件样式简单的衬衫来配那条裙子，不要样式太复杂的。
4. 他突然说了那么一大套理由，我不太相信他。

5. 这位国会女议员说她要辞职，但拒绝说明辞职的原因。
6. 她在新公司做同样的工作，但薪水比以前高。
7. 每年有一万人死于这种疾病——该数字相当于这个城镇的总人口。
8. 据说练瑜伽可以使人恢复内心的平静。
9. 英国队今年超水平发挥，打进了决赛。
10. 对于借款人来说，利率降低是个好消息。

# 3.5
# The procurement, leasing and maintenance of GA aircrafts
## 通用航空航空器选购、租赁和维修

## Background

### 1. Procurement of general aircrafts

Given the quite different user communities and the various properties and locations of flight operations, the selection of general aircrafts varies greatly. China covers a vast area of territories in which some underdeveloped regions are blocked and distanced by deserts and high mountains from the economically developed southeastern coastal areas. And over a long time, the resource advantages of China's western regions cannot be exerted because of the traffic hurdles, a bottleneck of the region's economic growth. In the western region of China, the railway is sparsely networked and the highway is of both low level and also of low density, and virtually most of towns and villages not accessible to highways are located there. Moreover, the airports in the regions are small, which makes it difficult for large airplanes to take off and land. All in all, users should purchase the right aircrafts based on their own geographical location, business needs and other actual situations.

Currently the Aviation Industry Corporation of China mainly produces small airplanes as Y-5, Y-12 and N5, transporters as Y-8, helicopters as Z-8, Z-9 and Z-11, and new products like Diamond, LE500 and MD600, which jointly play a great role in improving the transportation conditions in west China, strengthening ecological conservation, promoting the development of various industries as well as quickening the infrastructure construction.

As to foreign aircraft manufacturers like Bombardier, Cessna and Gulfstream, aircrafts manufactured by them are also optional, such as the business aircrafts, small rotorcrafts (like Brantly B-2B, Robinson 4, Cessna 172, Diamond DA-40 and Tiger), fixed-wing aircrafts, seaplanes and Mi-26 large helicopters.

Therefore, the decisions on the purchase or leasing of general aircrafts should be made after consultation with relevant parties and probably type proof by taking into consideration of various customer-specific factors, such as the purpose, location, frequency and method of aircraft application, the size of fleet as well as the geographical and weather information in the applied location.

### 2. Leasing of general aircrafts

Leasing is a transaction activity that under a signed contract the lessor collects the rent and the lessee pays the rent in exchange of the right to use lessor's assets. Specifically speaking, parties to the transaction activity enter into a contract under which the lessor shall assign the right to use the assets owned by the lessor to the lessee and collect the rent according to the contractual requirements, whereas the lessee shall be entitled to use such assets within a certain period of time and effect payment of rent to the lessor regularly. Parties to the leasing transaction shall fulfill their respective responsibilities and obligations under the contract and shall also be entitled to their corresponding rights. For lessor, his responsibilities are to transfer the equipment in compliance with the specifications to the lessee, and his rights are to collect rents in the specified amount on the specified date; and as to lessee, his responsibilities are to pay the rents to the lessor according to the specified requirements and his rights are to possess the right of use for the required equipment. In fact, through such leasing, the lessor obtains the right to use the required machines and equipment, through which he can then achieve the ultimate goal of fund raising as well.

Leases can be classified by different subject matters. In light of the nature of the lease, it can be classified as operating lease or financing lease; by the source of the lessor's assets, it can be divided into direct lease, sublease or sale and leaseback; according to the entitlement to preferential tax, the lease can be grouped into tax lease with entitlement to tax preference and sale lease without entitlement to tax preference. By the lessor's proportion of capital contribution in the purchase of a leased equipment, the lease can be typed as straight investment lease and leverage lease.

Aircraft leasing starts up late. In the early 1970s, leasing of used aircrafts began to appear and until 1980s, aircraft leasing made up 3% of the aircraft trading market only, because medium and large airlines at the time usually bought out the aircrafts directly. But later when great changes and fluctuations took place in the international oil price and the air transport market, airlines were forced to consider lowering their operational risks, and the high occupation of capital flow incurred in the purchase of aircrafts came out on the top of the problem list, so airlines turned to rent aircrafts rather than directly buy them, which empowered the aircraft leasing to be rapidly developed. From the late 1980s to the 1990s, the total number of airlines operated only with leased aircrafts went up from 15% to 49%, and in contrast, for those airlines operated only with bought aircrafts, it fell from 40% to 16%.

Aircraft leasing in the world's civil aviation industry is growing at robust pace and the aircraft leasing enterprises are taking up larger and larger shares in the aviation market. A dozen years ago, leasing companies claimed about 14% of the aircraft transportation market, and now many industry experts estimate that the claim has risen to 28%, meaning that more than 10,000 aircrafts out of the 36,000 commercial aircrafts operated across the world are leased.

### 3. Maintenance of general aircrafts

High-quality maintenance is, most importantly, for the purpose of ensuring flight safety, but the complexity of modern aircraft poses challenges on the aviation maintenance. The maintenance department undertakes to provide airworthy aircraft, and respond to customer requirements as much as possible under the reasonable budget, and the department should be responsible for the inspection, maintenance and repair of the aircraft to ensure the safe operation and maximum reliability of the aircraft.

With the continuous updating of GA technologies, the seemingly simple maintenance work can be at dead end, like when it is beyond the maintenance experience and level of technicians, or it necessitates special tools, or some maintenance procedures would naturally keep the aircraft out of service over a long run, etc. Therefore, it is inevitable to resort to external support for all maintenance in the whole life cycle of the aircraft. Today, the maintenance of general aircraft consists of outsourced maintenance and internal maintenance.

#### (1) Outsourced maintenance

The maintenance work may be carried out by two types of authorized contractors, i.e. individuals who have obtained technician licenses for aviation maintenance, as well as the legally established agencies or stations for aviation maintenance. The advantages of aviation maintenance stations are that their personnel's qualifications comply with the minimum requirements, and the supervision measures, quality control procedures as well as the high-level oversight from national administration authorizes all in place. For a certain work, technicians from the authorized maintenance agencies are usually more qualified and seasoned. And if the agencies are awarded an outsourcing contract for being good at maintaining that type of aircraft model, then logically they could provide a higher-level service. Prior to outsourcing any work, it is necessary to have a clear understanding of the type and scope of the work to be done, including the completion date of each separate procedure. In addition, a written agreement shall be concluded, which shall specify the maintenance provider's liability for work delay as well as the liable actions to be taken by him in case of any abnormal condition found during the maintenance work. Also, the completion progress of the work shall also be described in detail to determine the type and quantity of works performed and the parts used during the works. Finally, quality assurance shall be provided for the works performed and the parts used.

In addition, in the selection of maintenance service providers, their qualifications, reputation, work experience and price policies should be taken into account.

(2) Internal maintenance

As mentioned above, it is unlikely to complete all the maintenance work only with internal resources. For example, works like nondestructive tests, repair of parts, and the functional tests for a good many power plants and avionics are far beyond the capabilities of the majority of internal maintenance departments. However, people in such internal departments shall be competent to complete most of the internal scheduled maintenance and non-scheduled maintenance.

To conduct internal maintenance, the type and level of the maintenance work shall be determined. And if the maintenance is merely line maintenance or non-scheduled maintenance, or the work is low in level, it is undemanding for the technicians and their technical skills. But, if the work is not line maintenance or of high level, then the work should be transferred to the maintenance service providers who are technically qualified for the corresponding maintenance tasks.

In addition, the internal department may become more capable in maintenance work through seeking external information resources and consulting advices. On one hand, the manufacturer's representatives and their customer service personnel would usually render a wide range of services to assist customers. On the other hand, constant ties with some other technicians and staff working in the maintenance stations, especially those who have repaired the similar aircrafts would help customer get access to solutions to some technical problems from time to time. And the best part of internal maintenance lies in that when the needs for aircraft maintenance arise, the maintenance work in the aircraft can be overseen so as to ensure that the parts, technology and service are state-of-the-art.

## General Vocabulary

procurement   *n.* 采购；获得，取得
underdeveloped   *adj.* 不发达的
blocked   *adj.* 堵塞的；被封锁的
distanced   *v.* (使) 远离, 疏远；领先于 (一匹马) (distance 的过去式及过去分词)
desert   *n.* 沙漠, 荒原；沉闷乏味的境况 (或地区), 冷清的地方；应得的赏罚 (常用

复数）

hurdle   *vt.* 克服
sparsely   *adv.* 稀疏地；贫乏地
density   *n.* 密度
accessible   *adj.* 易接近的；可进入的；可理解的
strengthen   *vt.* 加强；巩固
ecological   *adj.* 生态的，生态学的
conservation   *n.* 保存，保持；保护
quicken   *vi.* 加快；变活跃；进入胎动期
optional   *adj.* 可选择的，随意的
rotorcraft   *n.* 旋翼飞机
fixed-wing   固定翼
seaplane   *n.* 水上飞机
consultation   *n.* 咨询；磋商；会诊；讨论会
frequency   *n.* 频率；频繁
leasing   *n.* 租赁，出租
lessor   *n.* 出租人
lessee   *n.* 承租人
asset   *n.* 资产；优点；有用的东西；有利条件；财产；有价值的人或物
assign   *vt.* 分配；指派；赋值
contractual   *adj.* 契约的，合同的
entitle   *v.* 给（某人）权利（或资格）；给……题名；称呼
fulfill   *vt.* 履行；实现；满足；使结束（与fulfil同）
responsibility   *n.* 责任，职责；义务
obligation   *n.* 义务；职责；债务
corresponding   *adj.* 相当的，相应的；一致的；通信的
compliance   *n.* 顺从，服从；符合；屈从；可塑性
specification   *n.* 规格；说明书；详述
specified   *adj.* 规定的；详细说明的
possess   *vt.* 控制；使掌握；持有；迷住；拥有，具备
ultimate   *adj.* 最终的；极限的；根本的
sublease   *n.* 转租
leaseback   *n.* 售后回租
entitlement   *n.* 权利；津贴
preferential   *adj.* 优先的；选择的；特惠的；先取的
leverage   *n.* 手段，影响力；杠杆作用；杠杆效率
bought out   买下……的全部产权
fluctuation   *n.* 起伏，波动
occupation   *n.* 职业；占有；消遣；占有期
incur   *v.* 招致，遭受；引致，带来……

empower　vt. 授权，允许；使能够

robust　adj. 强健的；健康的；粗野的；粗鲁的

dozen　n. 十二个，一打

complexity　n. 复杂，复杂性；复杂错综的事物

challenge　n. 挑战；怀疑

airworthy　adj. 适宜航空的；耐飞的；飞机性能良好的

reasonable　adj. 合理的，公道的；通情达理的

budget　n. 预算，预算费

seemingly　adv. 看来似乎；表面上看来

technician　n. 技师，技术员；技巧纯熟的人

inevitable　adj. 必然的，不可避免的

resort　vi. 求助，诉诸；常去；采取某手段或方法

external　adj. 外部的；表面的；外用的；外国的；外面的

outsource　vt. 把……外包

internal　adj. 内部的；体内的；（机构）内部的；国内的；本身的；内心的；（大学生）本校生的

authorized　adj. 经授权的；经认可的

qualification　n. 资格；条件；限制；赋予资格

comply　vi. 遵守；顺从，遵从；答应

oversight　n. 监督，照管；疏忽

seasoned　adj. 经验丰富的；老练的；调过味的

logically　adv. 逻辑上；合乎逻辑

prior to　在……之前；居先

conclude　vt. 推断；决定，作结论；结束

specify　vt. 指定；详细说明；列举；把……列入说明书

liability　n. 责任；债务；倾向；可能性；不利因素

abnormal　adj. 反常的，不规则的；变态的

assurance　n. 保证，担保；（人寿）保险；确信；断言；厚脸皮，无耻

reputation　n. 名声，名誉；声望

unlikely　adj. 不太可能的；没希望的

nondestructive　adj. 无损的；非破坏性的

power plant　发电厂；动力装置

avionic　adj. 航空电子学的；航电

competent　adj. 胜任的；有能力的；能干的；足够的

undemanding　adj. 要求不高的；容易的；不严格的

state-of-the-art　adj. 最先进的；已经发展的；达到最高水准的

## Notes

1. China covers a vast area of territories in which some underdeveloped regions are

blocked and distanced by deserts and high mountains from the economically developed southeastern coastal areas. And over a long time, the resource advantages of China's western regions cannot be exerted because of the traffic hurdles, a bottleneck of the region's economic growth.

- resource advantage　资源优势
- traffic hurdle　交通不便

译文：中国幅员辽阔，一些欠发达地区与经济发达的东南沿海之间，大漠为阻、高山相隔。西部地区的资源优势由于交通不便而长期得不到发挥，交通已成为制约西部地区经济发展的瓶颈。

2. Moreover, the airports in the regions are small, which makes it difficult for large airplanes to take off and land. All in all, users should purchase the right aircrafts based on their own geographical location, business needs and other actual situations.

- geographical location　地理位置

译文：西部地区机场小，不便大型飞机起降。因此用户要根据自身的地理位置、经营需要等实际情况选择购买飞机。

3. Currently the Aviation Industry Corporation of China mainly produces small airplanes as Y-5, Y-12 and N5, transporters as Y-8, helicopters as Z-8, Z-9 and Z-11, and new products like Diamond, LE500 and MD600, which jointly play a great role in improving the transportation conditions in west China, strengthening ecological conservation, promoting the development of various industries as well as quickening the infrastructure construction.

- Y-5　运-5（中国代号：Y-5，英文：Nanchang Y-5或Shijiazhuang Y-5）运输机，是20世纪50年代中国航空工业南昌飞机制造厂（代号：320厂）仿制生产的一型多用途单发双翼运输机。
- N5（又称农-5）系列农林飞机是中航工业洪都全面按照中国民用航空规章（CCAR23部适航条例）及型号合格审定程序研发的农林专用飞机。该机主要用于农作物的飞行作业、森林防火及农林业病虫害防治，经简单改装后还可以进行地质探测、空中摄影、航空体育训练和航空旅游等作业，具有使用成本低、适用范围广、作业能力强、操作性好、安全性高等特点。
- Z-8（中国代号：Z-8，英文：Changhe Z-8）直升机，是中国在20世纪90年代以法国SA321直升机（"超黄蜂"直升机）为基础仿制改进的13吨级多用途直升机，直-8直升机是单旋翼带尾桨多用途直升机，在标准状态下有较大的功率储备，具有飞行性能好、使用寿命长、飞行安全、操纵容易、使用维护方便、应急时可在水面起降等特点。
- transporter　运输机
- play a great role　发挥很大作用
- ecological conservation　生态保护

译文：目前，中国航空工业生产的小型飞机主要有运5、运12和农5，运输机有运8，直升机有直8、直9和直11，以及新产品"钻石""小鹰"500、MD600等，它们将在改善西部交通运输条件、加强生态环境保护、推动各个产业发展、加速基础设施建设等方面发挥巨大作用。

4. Therefore, the decisions on the purchase or leasing of general aircrafts should be made after consultation with relevant parties and probably type proof by taking into consideration of various customer-specific factors, such as the purpose, location, frequency and method of aircraft application, the size of fleet as well as the geographical and weather information in the applied location.

- take into consideration　顾及；考虑到……

译文：因此，通用飞机的购买或租赁，要根据用户的使用目的、使用地区、使用频率、使用方式和机队规模，以及使用地区的地理、气象等多方面的因素，向有关方面咨询，甚至进行机型论证后，再做出决策。

5. Specifically speaking, parties to the transaction activity enter into a contract under which the lessor shall assign the right to use the assets owned by the lessor to the lessee and collect the rent according to the contractual requirements, whereas the lessee shall be entitled to use such assets within a certain period of time and effect payment of rent to the lessor regularly.

- enter into a contract　与……订立契约，与……签订合同
- be entitled to　有权；有……的资格

译文：即交易各方签订合同，由出租人将其拥有的资产使用权转让给承租人并按照规定要求收取租金，承租人在一定时期内拥有该项资产的使用权并按期向出租人支付租金。

6. In light of the nature of the lease, it can be classified as operating lease or financing lease; by the source of the lessor's assets, it can be divided into direct lease, sublease or sale and leaseback.

- In light of　根据；鉴于；从……观点

译文：按租赁的性质可分为经营租赁与融资租赁；按出租人资产来源的不同可分为直接租赁、转租赁和售后回租等。

7. But later when great changes and fluctuations took place in the international oil price and the air transport market, airlines were forced to consider lowering their operational risks, and the high occupation of capital flow incurred in the purchase of aircrafts came out on the top of the problem list, so airlines turned to rent aircrafts rather than directly buy them, which empowered the aircraft leasing to be rapidly developed.

- be forced to　被迫做，被强迫做某事
- capital flow　资本流动；资金流量

译文：随后，国际市场油价的巨大变化及航空运输市场的巨大波动，使航空公司需要考虑降低公司运行风险，而购买飞机产生的占用资金流的问题是首先要考虑的因素，于是他们纷纷改买为租，从而使飞机租赁获得了快速发展。

8. Aircraft leasing in the world's civil aviation industry is growing at robust pace and the aircraft leasing enterprises are taking up larger and larger shares in the aviation market.

- take up　拿起；开始从事；占据（时间，地方）

译文：世界民航业的飞机租赁表现出强劲的势头。飞机租赁企业在航空市场上所占的比重

在逐渐上升。

9. The maintenance department undertakes to provide airworthy aircraft, and respond to customer requirements as much as possible under the reasonable budget, and the department should be responsible for the inspection, maintenance and repair of the aircraft to ensure the safe operation and maximum reliability of the aircraft.

— airworthy　适航性，是指航空器适合/适应于飞行（Fit to fly）的能力，是航空器的固有属性。适航性是通过航空器全寿命周期内的设计、制造、试验、使用、维护和管理的各个环节来实现和保持的。适航性要求首先体现在技术方面：系统安全性与物理完整性；其次体现在管理方面：技术状态与过程控制的管理等。

译文：维修部门的任务就是提供适航的飞机，并在合理成本下尽量有求必应，并负责检查、维护及修理飞机，以确保飞机的安全运行及最大程度的可靠性。

10. With the continuous updating of GA technologies, the seemingly simple maintenance work can be at dead end, like when it is beyond the maintenance experience and level of technicians, or it necessitates special tools, or some maintenance procedures would naturally keep the aircraft out of service over a long run, etc.

— dead end　死胡同，尽头

— out of service　失效；停止运行；故障；退役

译文：由于通用飞机各项技术的不断发展，看似简单的维修工作都有其限制因素，比如超出技术人员维修经验和水平、需要特殊工具或由于某些工序的性质让飞机长时间处于不可用的状态等。

11. The maintenance work may be carried out by two types of authorized contractors, i.e. individuals who have obtained technician licenses for aviation maintenance, as well as the legally established agencies or stations for aviation maintenance.

译文：承包商可以两种不同的授权方式进行维修：一是获得航空维修技术员执照的个人；二是通过批准成立的航空维修机构或航空维修站。

12. And if the agencies are awarded an outsourcing contract for being good at maintaining that type of aircraft model, then logically they could provide a higher-level service. Prior to outsourcing any work, it is necessary to have a clear understanding of the type and scope of the work to be done, including the completion date of each separate procedure.

译文：而且，如果该机构因擅长维修某种特定机型而被选中，应该可以提供较高水平的服务。在外包任何工作之前，必须对所要做的工作的类型和范围有一个清楚的了解，包括每个单独工序的完成日期。

## Exercises

Section one: Answer the following questions according to the text.
1. What are the differences between outsourced maintenance and internal

maintenance?

2. Why do people prefer the legally established agencies or stations for aviation maintenance?

3. What do we do if maintenance work beyond the maintenance experience and level of technicians?

4. Why and when did airlines consider lowering their operational risk and the occupation of capital flow in the purchase of aircrafts?

5. How can leases of aircrafts be classified by different subject matters?

**Section two: Match the words or phrases in column A with the definition in column B.**

| A | B |
| --- | --- |
| 1. lease | a. the quality of being intricate and compounded |
| 2. equipment | b. the largest possible quantity |
| 3. transaction | c. a region marked off for administrative or other purposes |
| 4. strengthen | d. a formal or official examination |
| 5. territory | e. property that is leased or rented out or let |
| 6. block | f. the act of transacting within or between groups |
| 7. maximum | g. render unsuitable for passage |
| 8. ultimate | h. an instrumentality needed for an undertaking or to perform a service |
| 9. inspection | i. make strong or stronger |
| 10. complexity | j. furthest or highest in degree or order; utmost or extreme |

**Section three: Translate the following into Chinese.**

1. The department should be responsible for the inspection, maintenance and repair of the aircraft to ensure the safe operation and maximum reliability of the aircraft.

2. Therefore, it is inevitable to resort to external support for all maintenance in the whole life cycle of the aircraft. Today, the maintenance of general aircraft consists of outsourced maintenance and internal maintenance.

3. For a certain work, technicians from the authorized maintenance agencies are usually more qualified and seasoned.

4. The completion progress of the work shall also be described in detail to determine the type and quantity of works performed and the parts used during the works. Finally, quality assurance shall be provided for the works performed and the parts used.

5. In addition, in the selection of maintenance service providers, their qualifications, reputation, work experience and price policies should be taken into account.

6. People in such internal departments shall be competent to complete most of the internal scheduled maintenance and non-scheduled maintenance.

7. If the work is not line maintenance or of high level, then the work should be transferred to the maintenance service providers who are technically qualified for the corresponding maintenance tasks.

8. On the other hand, constant ties with some other technicians and staff working in the maintenance stations, especially those who have repaired the similar aircrafts would help customer get access to solutions to some technical problems from time to time.

### Section four: Translate the following into English.

1. 确信公司的财务状况会大有改观。
2. 如果有孕在身,她就不该搬这些箱子。
3. 我要求我的学生守时。
4. 对于如何化解危机总统明显处于两难境地。
5. 律师为他们这个案子做了认真扎实的准备。
6. 这一发现是数年潜心研究的成果。
7. 国民医疗保健制度的改革将于明年实施。
8. 店主不应该向儿童出售小刀及其他锋利的器具。
9. 主人的姓名和地址都储存在芯片中,然后植入狗体内。
10. 他给他的孩子们灌输了一些奇怪的看法。

# Unit 4

# Applications of GA
通用航空的应用

Unit 4 Applications of GA 通用航空的应用

# 4.1 Emergency medical service
## 医疗救援

## Background

Air emergency medical service (EMS) is the flight service rendered for saving patient's life in time-critical situation and for conducting emergency rescue by using airplane or helicopter equipped with special medical care and rescue equipment.

### Air EMS abroad

Fixed-wing aircraft, as a basic guarantee, has been applied into air-based medical and rescue service several decades ago, and has established a full-fledged operation system in developed countries such as Europe and the USA. Taking the USA as an example, it is highly worthwhile for China to draw experience on its well-developed air rescue system.

In view of the principal implementer of air medical and rescue, the USA enjoys the strongest air rescue force in the world, which mainly includes Air Force, civil air patrol, state air rescue forces and other social search and rescue forces.

It is true that air ambulance service operators also account for a large part of the rescue forces. Statistics shows that at present there're more than 100 professional air ambulance service operators in the United States, which own above 1000 fixed-wing aircrafts (including business aircraft) for air medical treatment and ambulance, and conduct operations annually by over 700,000 times. Among them, Air Methods, the largest air medical transfer company in the United States, has a business presence covering 48 states of the country, owns 300 transfer service bases and 450 aircraft fleet (including helicopter and fix-wing), and executes air transfer operations annually by more than 100,000 times. And currently the Company is operated mainly in two models, one is community-based service, and the other is hospital-based service (HBS). Nevertheless, through a glance at how busy the air ambulance service operator is in the USA, it is easily understood that how well-developed the air rescue industry is in the country.

In view of the industry regulation and supervision, in the United States, the Federal Aviation Administration (FAA) formulates aircraft operation guidelines and policies, and also takes charge of certification and supervision of pilots and non-medical equipment. Furthermore, the industry shall adhere to the relevant provisions specified for air ambulance under the Federal Aviation Regulation No. FAR135 of the United States. With regard to the license of medical staff, assessment of professional nursing qualification, certification of medical equipment and the safety standard of medication in air EMS, the local medical and health authorities in the USA takes the responsibility of their supervision. And only if all above conditions are qualified, the Department of Health may issue the air ambulance permit.

In view of the cost and expenses, considering that the USA has a sound insurance coverage, the charges payable are in most cases settled by the patient's insurer even though each call for air ambulance service is quite pricy. The other approach is membership system. In such approach, members are entitled to be served once with air ambulance service at no cost of their own, provided that they pay a certain amount of membership fee on a yearly basis. Compared with these insurance plans that cover the air medical and ambulance service, the approach of membership system has less restrictions over the specific medical condition of the patient and members can apply for air ambulance service upon hospitalization.

At present, developed countries including USA, UK, Germany, France, Italy, Switzerland, Japan, Norway, Finland and Australia have all established a fairly sophisticated air emergency medical service system.

## Air EMS in China

In recent years, with the opening-up of low-altitude airspace and the continuous improvement of the air EMS system in China, helicopter-based air EMS has gradually come into the life of the public. Particularly as the parking apron and the platform for helicopter to conduct EMS is better established and developed, the air medical and rescue in China is serving the general public more and more as a kind of basic medical guarantee. And the public can call first-aid line 120 or 999 to ask for such air medical rescue. For some people, air EMS still looks far away from their life, but in times of critical condition, such way of medical and rescue can play an important role as the air EMS usually tackles certain special and urgent situations. Generally speaking, when serious diseases occur, such as cardiovascular and cerebrovascular diseases, poisoning, severe trauma, etc. and the ground ambulance cannot reach or cannot reach to the scene in a timely manner, people can apply for air medical rescue.

## Advantages of air EMS

1. Air EMS effectively improves the survival rate of patients. International medical statistics show that two-thirds of the seriously injured may die within 30 minutes. However, if the injured is attended with good medical care and treatment within 15 minutes, 80% of them would survive. Air medical rescue can effectively save medical time, improve the quality of medical treatment, and at the same time, it is conducive to future further medical treatment.

2. The safest way of medical rescue. Data validates that the accident rate of air ambulance rescue is far lower than that of ground ambulance rescue.

3. Air EMS also enjoys obvious efficiency advantages. According to a British study, the operating cost of an air ambulance is seven times higher than that of a ground ambulance, but its responsive service range is sixteen times wider than that of the latter.

Therefore, in order to cope with all kinds of complex disasters and catastrophes, and ensure the safety of people's lives and property, especially those frequently occurred small-scale accidents and medical rescue incidents, it is highly effective and necessary to apply helicopters and fixed wings to carry out medical rescue, which is timely, fast and less restricted by the geographical and traffic environment.

## Market prospects of air EMS

In the future, China's economic and social development will create strong demand for air medical rescue service. Apart from the non-normal emergency rescue such as earthquake relief, the normal demand mainly arises from four types: one is from the safeguard of major events, such as major sport events like the Winter Olympic Games, and people-gathering activities like important meetings; another one is from road traffic rescue. As there are more than 150 million car ownerships in China, and nearly 200,000 people die from traffic accidents every year, the demand for roadside assistance and rescue is growing. Third one is from special consumer group, such as Chinese expatriate staff, domestic and foreign tourists, foreign government officials posted in China, enterprise executives, etc. Last one is from the guarantee of mass public's wellbeing and health, like the case of sudden attack of serious and critical medical condition in everyday life.

## General Vocabulary

time-critical   *adj.* 时序要求严格的
guarantee   *n.* 保证；担保；保证人；保证书；抵押品
worthwhile   *adj.* 重要的；令人愉快的；有趣的；值得（花时间、金钱、努力等）的；有价值的，有益的；值得做的
implementer   *n.* 实施者；制订人；实现器；实作器
ambulance   *n.* 救护车；战时流动医院
execute   *vt.* 实行；执行；处死
glance   *n.* 一瞥；一滑；闪光；（板球）斜击；辉金属
well-developed   *adj.* 发达的；发育良好的
regulation   *n.* 管理；规则；校准
guideline   *n.* 指导方针
certification   *n.* 证明，鉴定；出具课程结业证书，颁发证书
adhere   *vi.* 坚持；依附；黏着；追随
with regard to   关于；至于
assessment   *n.* 评定；估价
permit   *n.* 许可证，执照
insurance   *n.* 保险；保险费；保险契约；赔偿金
coverage   *n.* 覆盖，覆盖范围；新闻报道；保险范围
insurer   *n.* 保险公司；承保人
pricy   *adj.* 价格高的；昂贵的
membership   *n.* 资格；成员资格；会员身份
restriction   *n.* 限制；约束；束缚
hospitalization   *n.* 住院治疗；医院收容；住院保险（与 hospitalization insurance 同）
Switzerland   *n.* 瑞士（欧洲国家）
Norway   *n.* 挪威（北欧国家名）
sophisticated   *adj.* 复杂的；精致的；久经世故的；富有经验的
platform   *n.* 平台；月台，站台；坛；讲台
first-aid   *adj.* 急救用的
tackle   *v.* 应付，处理（难题或局面）；与某人交涉；（足球、曲棍球等）抢球；（橄榄球或美式足球）擒抱摔倒；抓获；对付，打（尤指罪犯）

cardiovascular   *adj.* 心血管的
cerebrovascular   *adj.* 脑血管的
poisoning   *n.* 中毒
trauma   *n.* 创伤（由心理创伤造成精神上的异常）；外伤
timely   *adj.* 及时的；适时的
survival   *n.* 幸存，残存；幸存者，残存物
injure   *vt.* 伤害，损害
treatment   *n.* 治疗，疗法；处理；对待
conducive   *adj.* 有益的；有助于……的
validate   *vt.* 证实，验证；确认；使生效
responsive   *adj.* 响应的；应答的；回答的
latter   *adj.* 后者的；近来的；后面的；较后的
disaster   *n.* 灾难，灾祸；不幸
catastrophe   *n.* 大灾难；大祸；惨败
small-scale   *adj.* 小规模的
incident   *n.* 事件，事变；插曲
prospect   *n.* 前途；预期；景色
non-normal   非正常，非常规的
arise   *vi.* 出现；上升；起立
safeguard   *n.* 保护；保卫；保护措施；预防措施
winter olympic games   冬季奥运会
roadside   *adj.* 路边的；路旁的
expatriate   *n.* 移居国外者，侨民；被流放者
tourist   *n.* 旅行者，观光客
official   *n.* 官员；公务员；高级职员
executive   *n.* 主管；行政领导；（政府的）行政部门；执行委员会
wellbeing   *n.* 幸福；福利；安乐
sudden   *adj.* 突然的，意外的；快速的

# Notes

1. Air emergency medical service (EMS) is the flight service rendered for saving patient's life in time-critical situation and for conducting emergency rescue by using airplane or helicopter equipped with special medical care and rescue equipment.
— emergency medical service (EMS)  紧急医疗服务，也称"急救"。当有任何意外或急病发生时，施救者在医护人员到达前，按医学护理的原则，利用现场适用物资临时及适当地为伤病者进行的初步救援及护理，然后从速送院。
译文：航空应急医疗救援，就是使用装有专用医疗救护设备的飞机或直升机，为抢救患者生命和紧急施救进行的飞行服务。
2. Fixed-wing aircraft, as a basic guarantee, has been applied into air-based

medical and rescue service several decades ago, and has established a full-fledged operation system in developed countries such as Europe and the USA.

— fixed-wing aircraft　固定翼飞机，是指由动力装置产生前进的推力或拉力，由机身的固定机翼产生升力，在大气层内飞行的重于空气的航空器。它是固定翼航空器的一种，也是最常见的一种，另一种固定翼航空器是滑翔机。飞机按照其使用的发动机类型又可被分为喷气飞机和螺旋桨飞机。

— developed country　发达国家

译文：作为一种基本保障，早在几十年前固定翼飞机就已应用于航空医疗救援，并在欧美等发达国家拥有成熟的运营体系。

3. Among them, Air Methods, the largest air medical transfer company in the United States, has a business presence covering 48 states of the country, owns 300 transfer service bases and 450 aircraft fleet (including helicopter and fix-wing), and executes air transfer operations annually by more than 100,000 times. And currently the Company is operated mainly in two models, one is community-based service, and the other is hospital-based service (HBS).

— Air Methods　创立于1980年，总部位于美国科罗拉多州Englewood，全职雇员4554人，连同其子公司，在美国提供直升机运营，提供全美航空医疗紧急运送服务给各家医院。以独立方式提供航空医疗运输服务于一般民众、医院及签有独家经营合作协议的机构。服务包括飞机的操作和维修、医疗保健、派遣和通讯、医疗计费和收款。截至2015年12月31日，该部门的飞机机队包括321架公司自有的飞机和92架租赁飞机，以及54架客户的飞机，由AMS与他们签订合同经营。

译文：其中美国最大的空中医疗转运公司Air Methods的业务覆盖美国48个州，共有300个转运服务基地，机队规模达450架（包括直升机和固定翼），每年空中转运10万次以上。该公司目前主要营运两种模式：一种是基于社区的服务，一种是基于医院的服务（HBS）。

4. With regard to the license of medical staff, assessment of professional nursing qualification, certification of medical equipment and the safety standard of medication in air EMS, the local medical and health authorities in the USA takes the responsibility of their supervision. And only if all above conditions are qualified, the Department of Health may issue the air ambulance permit.

— takes responsibility of　负责，承担责任

— Department of Health　卫生部，美国卫生与公众服务部（United States Department of Health and Human Services；HHS），又译美国健康及人类服务部或美国卫生部，是维护美国公民健康，提供公共服务的联邦政府行政部门，其中一项主要工作就是为没有能力治疗的人群提供医保服务。

— air ambulance　空中救护

译文：而医务人员的从业许可、专业护理水平考核、医疗装备认证及航空用药的安全标准则由美国当地的卫生部门负责监管。只有在以上所有项目合格的情况下，卫生署才会颁发"空中救护许可证"。

5. Compared with these insurance plans that cover the air medical and ambulance service, the approach of membership system has less restrictions over the specific

medical condition of the patient and members can apply for air ambulance service upon hospitalization.

- compared with　与……相比，和……比起来，与……相比较
- membership system　会员制
- apply for　申请，请求

译文：相较于含有空中医疗救护服务的保险产品，会员制的产品对于病患的病情没有严格的限制，会员在住院后便可申请空中医疗救护服务。

6. In recent years, with the opening-up of low-altitude airspace and the continuous improvement of the air EMS system in China, helicopter-based air EMS has gradually come into the life of the public. Particularly as the parking apron and the platform for helicopter to conduct EMS are better established and developed, the air medical and rescue in China is serving the general public more and more as a kind of basic medical guarantee.

- low-altitude airspace　低空空域
- parking apron　停机坪
- more and more　越来越

译文：近年来，伴随中国低空空域开放及航空医疗救援体系的不断完善，直升机为主体参与的航空医疗救援逐步进入了大众的视野。特别是随着停机坪、应急医疗救援（EMS）直升机到医疗救援平台的不断搭建和发展，我国的航空医疗救援作为一种基本医疗保障，离百姓生活越来越近。

7. Generally speaking, when serious diseases occur, such as cardiovascular and cerebrovascular diseases, poisoning, severe trauma, etc. and the ground ambulance cannot reach or cannot reach to the scene in a timely manner, people can apply for air medical rescue.

- Generally speaking　一般来说，一般而言
- in a timely manner　及时，在一个及时的方式

译文：通常来说，当发生重症疾病时，如心脑血管疾病、中毒、严重创伤等，在地面救护车无法到达或无法快速到达现场时，人们就可以申请航空医疗救援近。

8. International medical statistics show that two-thirds of the seriously injured may die within 30 minutes. However, if the injured is attended with good medical care and treatment within 15 minutes, 80% of them would survive.

译文：国际医疗统计表明，重伤者中，2/3的人会在30分钟之内死亡。如果受伤者在15分钟内得到良好的救护和治疗，80%的人会保住生命。

9. Therefore, in order to cope with all kinds of complex disasters and catastrophes, and ensure the safety of people's lives and property, especially those frequently occurred small-scale accidents and medical rescue incidents, it is highly effective and necessary to apply helicopters and fixed wings to carry out medical rescue, which is timely, fast and less restricted by the geographical and traffic environment.

- in order to　为了，以便，目的，为着
- cope with　处理，应付

译文：所以，为应对错综复杂的各类灾害和灾难，保证人民生命财产安全，特别是日常频发的小规模事故灾难和医疗救助事件，运用直升机、固定翼这种及时、快速、较少受地域和交通环境限制的救援运输工具开展医疗救护是十分有效和必要的。

10. Third one is from special consumer group, such as Chinese expatriate staff, domestic and foreign tourists, foreign government officials posted in China, enterprise executives, etc. Last one is from the guarantee of mass public's wellbeing and health, like the case of sudden attack of serious and critical medical condition in everyday life.

— expatriate staff　离国服务的工作人员
— government officials　政府官员，公务人员

译文：三是特殊消费群体类，比如中国驻外人员、境内外游客以及各国政府驻华人员、企业高管等；四是民众民生健康保障类，日常生活中的突发重危病情形。

## Exercises

Section one: Answer the following questions according to the text.
1. What is air emergency medical service (EMS)?
2. Why the American air rescue system is highly worthwhile for China to draw experience from?
3. In view of the industry regulation and supervision, what is the Federal Aviation Administration (FAA) in the US responsible for?
4. In view of the cost and expenses, what are two types of insurance coverage of air rescue service in United States?
5. What is the current situation of air EMS in China?

Section two: Match the words or phrases in column A with the definition in column B.

A　　　　　　　　　　B
1. platform　　　　　a. a mishap; especially one causing injury or death
2. disease　　　　　 b. a thorough physical examination
3. poisoning　　　　c. a contented state of being happy and healthy and prosperous
4. ambulance　　　 d. an event resulting in great loss and misfortune
5. treatment　　　　e. the possibility of future success
6. disaster　　　　　f. a raised horizontal surface
7. prospect　　　　 g. an impairment of health or a condition of abnormal functioning
8. medical　　　　  h. the physiological state produced by a poison or other toxic substance
9. wellbeing　　　　i. a vehicle that takes people to and from hospitals

10. accident                 j. care by procedures

## Section three: Translate the following into Chinese.

1. Air emergency medical service (EMS) is the flight service rendered for saving patient's life in time-critical situation and for conducting emergency rescue by using airplane or helicopter equipped with special medical care and rescue equipment.

2. Statistics shows that at present there're more than 100 professional air ambulance service operators in the United States, which own above 1000 fixed-wing aircrafts (including business aircraft) for air medical treatment and ambulance, and conduct operations annually by over 700,000 times.

3. Furthermore, the industry shall adhere to the relevant provisions specified for air ambulance under the Federal Aviation Regulation No. FAR135 of the United States.

4. At present, developed countries including USA, UK, Germany, France, Italy, Switzerland, Japan, Norway, Finland and Australia have all established a fairly sophisticated air emergency medical service system.

5. In recent years, with the opening-up of low-altitude airspace and the continuous improvement of the air EMS system in China, helicopter-based air EMS has gradually come into the life of the public.

6. For some people, air EMS still looks far away from their life, but in times of critical condition, such way of medical and rescue can play an important role as the air EMS usually tackles certain special and urgent situations.

7. Air medical rescue can effectively save medical time, improve the quality of medical treatment, and at the same time, it is conducive to future further medical treatment.

8. The safest way of medical rescue. Data validates that the accident rate of air ambulance rescue is far lower than that of ground ambulance rescue.

## Section four: Translate the following into English.

1. 影片的整个情节荒谬得令人难以置信。
2. 她指责了该政党，也含蓄地指责了该党的领袖。
3. 公司正在削减开支，真不知道这将会对我们部门产生什么影响。
4. 他将她的评论理解为对政府的含蓄批评。
5. 商店说会给我们退换那台电视机，因为它还没有过包退期。
6. 这种面包的标签上写着保证不含防腐剂。
7. 这台冰箱保修期是3年。
8. 在这些问题上我总是依靠父亲指点迷津。
9. 那个繁忙的车辆入口处对行人来说很不安全。
10. 自从开始这种新的饮食，她的健康状况大为改善。

# 4.2 Agricultural aviation
## 农业航空

## Background

### 1. Concept of agricultural and forestry aviation

Agricultural and forestry aviation, a unique term in the field of general aviation, generally includes agricultural aviation, aerial forest fire-protection and artificial precipitation in which the agricultural aviation is in the dominant place and has a relatively steady growth. Agricultural aviation is an operational flight work involved in the production and disaster relief of sectors of agriculture, forestry, animal husbandry and fishery by using civil aircrafts, such as crop's chemical weeding, foliage dressing, spraying of microelements, insect pest and disease control, grassland's seeding, forest pest control and other flight engagements. Agricultural aviation can greatly improve labor productivity and play an important role in the prevention of natural disasters, pest control, improvement of human living environment and ecological balance. In particular, for the vast open agricultural lands in the northern China, aviation operation is a necessary condition to ensure the true and more efficient implementation of mass production in the lands.

### 2. Application of agricultural and forestry aviation operation

The main roles that agricultural aviation plays are as follows:

(1) Agricultural aviation operations can improve the quality of field management and ensure stable and high yields.

The consistent high and stable yield of crops depends on the increasingly higher management skills, whereas currently the poor field management in the middle and later stages of crops stands as the main issue in China's agricultural production. It is in the middle and later growth period of crops that technical measures should be carried out, which includes control of plant diseases and insect pests, foliage spraying of microelements, foliage dressing, promotion of early maturity of crops, plant care and disease prevention, and quality improvement of crops. However, during this period, the crops grow vigorously, making the cropland lush and thick and accordingly the labor works painstaking and inefficient. Also, machine works in the cropland could easily trip or overwhelm the plants and may even bruise the plants' root system and more importantly, it is difficult to realize a uniform and rapid control if there're vast areas of fields. Nevertheless, applying aircraft operations will get rid of above dilemma, because agricultural aviation operations, being irreplaceable by ground-based machinery and manpower, can achieve a good effect of high quality in field management and meanwhile does no harm to the physical structure of soil as well as the later growth of crops. Moreover, since the ground-based machine works are vulnerable to the impact of weather conditions, especially in the severe water-logging seasons, it turns out that only the aircraft can give its unique play in the implementation of agricultural works at such time.

(2) Agricultural aviation operations are better at taking the field's situation by surprise and at preventing and controlling major biological disasters.

The sudden biological disaster is characterized with intensive outbreak, fast spread and drastic hazards. Therefore both the affected farmlands and those non-cultivated public areas should be controlled immediately and uniformly so as to fundamentally curb its harms. Pests like grass moth, East Asian migratory locust and army worm have all wreaked havoc due to the essential slowness of manual control. However, agrochemical dispersing operations by aircraft can take a strong, efficient and effective control of the plant diseases and insect pests by surprise as working above the field with aircraft is highly maneuverable and has a wide range of working scope. Particularly in the rainy years with severe floods in the spring and summer, the superior role of agricultural aviation operations draws more attention.

(3) Agricultural aviation operations are highly efficient and can help achieve mass production of lands in steps.

Agrochemical dispersing operations by aircraft are efficient, fast and effective in disease and pest control. Also, as it applies one tenth as many amounts of chemicals as manual spraying at best, it is low in costs. For example, in the growth period of rice, chemicals should be applied for 2 or 3 times, with a dosage of 500ml chemicals plus a labor cost of about 35 yuan per arce in each manual application. However, if the chemicals are dispersed by aircraft operation, it takes only a dosage of 100ml chemicals and an operation cost of 6-8 yuan per arce per dispersing, which by contrast, reduces chemical dosage by 80% and saves labor cost by 58 up to 81 yuan per arce. Huge economic benefits could thus be created.

(4) Agricultural aviation operations can save costs, increase efficiency and lower pesticide residues.

Through ultra-micro dispersing practice of chemicals by aircraft, the dispersing work is done rapidly and the control is all-sided as the plant's leaves are whipped up with the flow of air created by the flying of aircrafts so that both sides of leaves can be coated with chemicals or fertilizer. Moreover, given that the chemicals are sprinkled evenly in the form of tiny droplets, they can be effectively attached to the surface of plants and pests for absorption and then the pathogens and pests in plants could be killed, meanwhile, the sprinkled chemicals can also wipe out the moving pathogens and pests in the air, which will comprehensively hold down the infection of pathogens and keep control of its spread. In this case, the control and prevention effect can be improved by 15% up to 35%. And if during the key growth stages of crops, fertilizer and microelements are dispersed by aircrafts on their leaves instead of being simply tossed out by hands or mixed into the soil, the yields can be significantly increased on average by over 10%

Artificial precipitation refers to fly aircrafts in the cloud to disturb and stir the air and spray catalysts to promote the rapid formation of ice crystals or cloud and fog droplets in the clouds so as to get rainfall and snowfall when the air condition for precipitation is not adequate. The precipitation can be increased by 10%~20% if catalysts are dispersed on deep stratiform clouds and by 100%~200% if on strong cumulus clouds. Apart from alleviating drought in the agricultural and pastoral areas, artificial precipitation can also reduce the fire risk and help extinguish fires in forest regions.

Aerial forest fire-protection is to prevent and fight forest fires with aviation technologies. It generally includes daily patrol, fire extinguishing by airlanding and aerial firefighting. For the latter two manners, they are usually operated with large helicopters, while for the daily patrol, it is mostly performed with low-cost fixed-wing aircrafts, and recently the real-time image monitoring of UAVs is also alternatively applied for that end. Nevertheless, in China, the forest coverage is relatively low; therefore the workloads of aerial forest fire-protection remain relatively alike in each year.

### 3. Development of agricultural and forestry aviation

Aircrafts, if assembled with sprinkling system, can be used for pest control, fertilization and weeding for forests, grasslands and crops, and in another way, if assembled with seeding system, can conduct aerosowing for forestation, grass planting and seeding of rice and soybean as well as protect forest from fires. As early as 1911, Germans proposed to spray chemicals with airplanes to control forest pests. And in 1918, the United States succeeded in controlling forage pests through dispersing chemicals by aircrafts and then countries like Canada and the Soviet Union all applied aircrafts to agriculture one after another.

After World War II, a variety of insecticides, fungicides and herbicides sprang up in the market which necessitated the use of efficient spraying machines and tools to keep up with the work requirements. At the same time, large numbers of military aircrafts and pilots were transferred to agricultural sector, paving the way to the booming of agricultural and forestry aviation. In 1950s, special-purpose and multi-purpose aircrafts dedicated for agricultural application were designed and manufactured for the first time. And then in late 1950s, helicopters joined agricultural aviation. Up to 1983, it was calculated around the world that there were about 32,000 agricultural aircrafts and a total of operation areas of 5.6 billion acres which approximately accounted for 25% of total sown area worldwide.

Over the 60 years from 1952, agricultural and forestry aviation in China has made great progress. In 2013, Chinese agricultural and forestry aviation was recorded with flight operations of $3.2 \times 10^4$h (equal to 6% of the total GA operational flight hours) and agricultural aerial dispersing areas of $285 \times 10^4 hm^2$. Besides, its applications were also seen to expand from a single locust control to many other fields, such as prevention of rice and wheat diseases, integrated control of fall webworms and drought-related insects, grassland seeding, rodent control and forest firefighting. The agricultural and forestry aviation is one of the most important parts of general aviation industry in China, but, as limited by the small scale and scattered operation of agricultural operators in China, it is in a sustained and stable but low growth.

By the end of 2016, of the 189 GA enterprises in operation in China, 129 enterprises are qualified for flight operations, and 42 enterprises are qualified for agricultural and forestry operations out of which 37 enterprises are in normal operation. Agricultural and forestry operations by aircraft have become one of the key business markets of China's GA enterprises.

## 4. Application of agricultural and forestry aircrafts

Nowadays, agricultural and forestry aircrafts are playing a huge role in the agricultural production of various countries, and with the advancement of aviation technology, their technical performance, application scope and operation capacity will continue to improve, and the agricultural production in different countries in the world will be more modernized. And as the average service life of agricultural and forestry aircraft is about 30-35 years, before long, a large number of agricultural and forestry aircrafts produced in 1970-1980 will be phased out. Hence, it is estimated in the following decade that the manufacture and sales of worldwide agricultural and forestry aircrafts will embrace a steady and solid growth.

In addition, agricultural and forestry aircrafts are applied elsewhere. They, with the use of remote sensing technology, can assist in spotting weeds and managing water shortage for crops and in preventing and fighting fires. In the United States, a single engine aerial tanker (SEAT), a kind of refitted or manufactured agricultural aircraft, is often used to spray water, flame retardants or inhibitors on wildfires. Some local or state fire-fighting agencies in America have deployed SEAT to carry out fire extinguishing operations by more than 15 years, and for the federal government, it is since 1950s when they have begun to use such aircraft.

(1) GA-200 Fatman

GA-200 Fatman is a two-seater agricultural and forestry aircraft developed by Australia-based Gippsland Aeronautics Co., Ltd.. Founded in 1971, the company is mainly specialized in the modification business of aircrafts including wooden aircraft and turboprop aircraft, and it also develops GA-200 Fatman and GA-8 AirVan on its own. The GA-200 prototype aircraft was registered in 1991 and obtained general-purpose and agricultural type certificate from Australian Civil Aviation Safety Authority (CASA) on March 1$^{st}$, 1991. As the aircraft complies with Australian CASA Standard No. 101.16 and No. 101.22 as well as the Federal Aviation Regulations (FAR) No. 23 and No.36 Amendments of the United States, in October 1997, it obtained restricted type certificate according to the FAR No.23 standards. In total, 45 such aircrafts were delivered up to February 2003.

(2) EMB-202 Ipanema

EMB-202 Ipanema is a small single-seater agricultural and forestry aircraft designed by Embraer S.A. and manufactured by Nevada Aviation Industry (see Figure 6-2). Nevada Aviation Industry was founded in 1954 and then turned into a wholly-owned subsidiary of Embraer through acquisition on March 10, 1980. Since May, 1969, EMB-202 began its design work and its prototype aircraft EMB-200 was put on maiden flight on July 30, 1970. The agrochemical container of the aircraft can be loaded with 950L liquid chemical agent or 750kg solid chemical agent. With the spray system installed under the middle of the fuselage, the spray boom or Miconier's nebulizers are mounted respectively behind and above the trailing edge of the wing. Moreover, it is optional to assemble with ram air compressor, light spray boom, Miconier's AU5000 rotary spray system, trapezoidal distributor of solid chemicals with variable nozzle width as well as some electronics including portable VHF transceiver and Garmin's portable GPS receiver.

## General Vocabulary

dominant　*adj.* 显性的；占优势的；支配的，统治的
crop　*n.* 产量；农作物；庄稼；平头
chemical　*n.* 化学制品，化学药品
weeding　*n.* 除草，除杂草
foliage　*n.* 植物；叶子（总称）
dressing　*n.* （拌制色拉用的）调料；馅，填料；（保护伤口的）敷料；穿戴；（精修织物用的）浆料；肥料
microelement　*n.* 微量元素
insect　*n.* 昆虫；卑鄙的人
pest　*n.* 害虫；有害之物；讨厌的人
seeding　*n.* 播种；晶种
July　*n.* 七月
kill　*vt.* 杀死；扼杀；使终止；抵消
engagement　*n.* 婚约；约会；交战；诺言；进场（游戏术语）
ecological　*adj.* 生态的，生态学的
join　*vt.* 参加；结合；连接
implementation　*n.* 实现；履行；安装启用
yield　*n.* 产量；利润，红利率
early maturity　早熟
cropland　*n.* 农田；植作物之农地
lush　*adj.* 丰富的，豪华的；苍翠繁茂的
painstaking　*adj.* 艰苦的；勤勉的；小心的
trip　*vi.* 绊倒，跌倒；轻快地走（或跑、跳舞）；（非正式）（服用毒品后）产生幻觉；（部分电路）自动断开；作短途旅行
overwhelm　*vt.* 淹没；压倒；受打击；覆盖；压垮
bruise　*vt.* 使受瘀伤；使受挫伤

uniform  *adj.* 统一的；一致的；相同的；均衡的；始终如一的
dilemma  *n.* 困境；进退两难；两刀论法
manpower  *n.* 人力；人力资源；劳动力
vulnerable  *adj.* 易受攻击的，易受……的攻击；易受伤害的；有弱点的
impact  *n.* 影响；效果；碰撞；冲击力
severe  *adj.* 严峻的；严厉的；剧烈的；苛刻的
water-logging  水浸，水渗；浸透
biological  *adj.* 生物的；生物学的
characterize  *vt.* 描绘……的特性；具有……的特征
intensive  *adj.* 加强的；集中的；透彻的；加强语气的
drastic  *adj.* 激烈的；猛烈的
hazard  *n.* 危险，风险；冒险的事；机会；双骰子游戏；（高尔夫球的）球场障碍；（庭院网球用语）可得分区域；（台球）落入袋中
cultivated  *adj.* 用于耕种的；有教养的；栽培的，培植的
immediately  *adv.* 立即，立刻；直接地
uniformly  *adv.* 一致地
fundamentally  *adv.* 根本地，从根本上；基础地
curb  *n.* 抑制；路边；勒马绳
grass moth  草地蛾
east Asian migratory locust  东亚飞蝗
army worm  黏虫
wreak  *vt.* 发泄；报仇；造成（巨大的破坏或伤害）
havoc  *n.* 大破坏；浩劫；蹂躏
dispersing  *n.* 分配，分散
maneuverable  *adj.* 有机动性的；容易操作的；可调动的
dosage  *n.* 剂量，用量
pesticide  *n.* 杀虫剂
residue  *n.* 残渣；剩余；滤渣
ultra-micro  超微量
all-sided  *adj.* 全面的
whip up  激起；鞭打
coat  *vt.* 覆盖……的表面
fertilizer  *n.* 肥料；受精媒介物；促进发展者
evenly  *adv.* 均匀地；平衡地；平坦地；平等地
droplet  *n.* 小滴，微滴
absorption  *n.* 吸收；全神贯注，专心致志
pathogen  *n.* 病原体；病菌
comprehensively  *adv.* 包括地；包括一切地
infection  *n.* 感染；传染；影响；传染病
toss out  *vt.* 扔出

disturb　*vt.* 打扰；妨碍；使不安；弄乱；使恼怒
stir　*v.* 搅拌；微动；（使）活动；激发，打动；（非正式）挑拨；传播
catalyst　*n.* 催化剂；刺激因素
formation　*n.* 形成；构造；编队
crystal　*n.* 结晶，晶体；水晶；水晶饰品
stratiform　*adj.* 分层排列的；（矿床）成层的
cumulus　*n.* 积云；堆积，堆积物
alleviate　*vt.* 减轻，缓和
drought　*n.* 干旱；缺乏
pastoral　*adj.* 牧师的；牧人的；田园生活的；乡村的
extinguish　*vt.* 熄灭；压制；偿清
airlanding　*v.* （军队或物资等）用飞机降落
real-time　*adj.* 实时的；接到指示立即执行的
UAV　*abbr.* 无人机（unmanned aerial vehicle）
aerosowing　飞机播种
forestation　*n.* 造林；森林管理
forage　*n.* 饲料；草料；搜索
Soviet Union　苏联［1922—1991年，首都莫斯科（Moscow），位于欧、亚洲］
insecticide　*n.* 杀虫剂
fungicide　*n.* 杀真菌剂
herbicide　*n.* 除草剂
spring up　出现；涌现；萌芽
fall webworms　美国白蛾
rodent　*n.* 啮齿目动物（如老鼠等）
scattered　*adj.* 分散的；散乱的
modernized　*adj.* 现代化的
phase out　使逐步淘汰；逐渐停止
embrace　*vt.* 拥抱；信奉，皈依；包含
elsewhere　*adv.* 在别处；到别处
sensing　*n.* 感觉，察觉
spotting　*v.* 发现；赏识；（尤指体育、娱乐业）业余观察；（尤指从空中）确定敌人的位置；沾上污渍；生斑；散布；用点装饰；下零星小雨；将（台球）放在置球点上；（比赛或运动中）让……一步；借（钱）给（spot 的现在分词）
tanker　*n.* 油轮；运油飞机；油槽车；坦克手
retardant　*n.* 阻滞剂；抑止剂
inhibitor　*n.* 抑制剂，抗化剂；抑制者
wildfire　*n.* 野火，烈火；希腊火；高度易燃物质，燃烧剂；鬼火
aeronautics　*n.* 航空学；飞行术
modification　*n.* 修改，修正；改变
prototype　*n.* 原型；标准，模范

amendment　*n.* 修正案；改善；改正
Embraer S.A.　巴西航空工业公司
chemical agent　化学剂；化学药剂
nebulizer　*n.* 喷雾器
compressor　*n.* 压缩机；压缩物；收缩肌；压迫器
trapezoidal　*adj.* 梯形的；不规则四边形的
nozzle　*n.* 喷嘴；管口；鼻
portable　*adj.* 手提的，便携式的；轻便的
transceiver　*n.* 收发器，无线电收发两用机
GPS　*abbr.* 全球定位系统（Global Position System）

## Notes

1. Agricultural aviation is an operational flight work involved in the production and disaster relief of sectors of agriculture, forestry, animal husbandry and fishery by using civil aircrafts, such as crop's chemical weeding, foliage dressing, spraying of microelements, insect pest and disease control, grassland's seeding, forest pest control and other flight engagements.
　— chemical weeding　化学除草
　— insect pest　虫害
　— foliage dressing　叶面施肥
　译文：农业航空（agricultural aviation），是使用民用航空器从事农业、林业、牧业、渔业生产及抢险救灾的作业飞行，如农作物化学除草、叶面施肥、喷施微量元素、防治病虫害、草原播种、防治森林害虫等飞行活动。

2. However, during this period, the crops grow vigorously, making the cropland lush and thick and accordingly the labor works painstaking and inefficient. Also, machine works in the cropland could easily trip or overwhelm the plants and may even bruise the plants' root system and more importantly, it is difficult to realize a uniform and rapid control if there're vast areas of fields.
　译文：但该时期作物长势繁茂，田间郁闭，劳动强度大，工作效率低，机械作业极易躺倒或压倒植株，还可能压伤植物根系，在一个较大的区域内很难达到统一防治和迅速控制的目的。

3. Moreover, since the ground-based machine works are vulnerable to the impact of weather conditions, especially in the severe water-logging seasons, it turns out that only the aircraft can give its unique play in the implementation of agricultural works at such time.
　— vulnerable to　易受……的攻击；易受……的侵害
　译文：由于受气候条件影响，尤其是遇到内涝严重季节，地面机械无法进行作业，此时唯有飞机可以发挥它的独特性能，实现农事操作。

4. However, agrochemical dispersing operations by aircraft can take a strong,

efficient and effective control of the plant diseases and insect pests by surprise as working above the field with aircraft is highly maneuverable and has a wide range of working scope. Particularly in the rainy years with severe floods in the spring and summer, the superior role of agricultural aviation operations draws more attention.

- agrochemical dispersing 农用化学品喷洒
- working scope 作业半径

译文：采用航化作业，飞机作业机动性强，作业半径大，在防治农作物病虫害方面显示了很强的突击能力。同时，飞机作业效率高，作业效果好，在严重春涝、夏涝的雨年份更能显示其优越性。

5. Through ultra-micro dispersing practice of chemicals by aircraft, the dispersing work is done rapidly and the control is all-sided as the plant's leaves are whipped up with the flow of air created by the flying of aircrafts so that both sides of leaves can be coated with chemicals or fertilizer.

- ultra-micro dispersing 超微量喷洒
- whip up 激起；鞭打
- be coated with 用……包上；表面覆盖有……；表面镀……

译文：飞机航化作业采用超微量喷洒法，作业时间短，还可以实现立体防治，即利用飞机飞行产生的气流吹动植物叶片，使叶片正反面均能着药着肥。

6. Artificial precipitation refers to fly aircrafts in the cloud to disturb and stir the air and spray catalysts to promote the rapid formation of ice crystals or cloud and fog droplets in the clouds so as to get rainfall and snowfall when the air condition for precipitation is not adequate.

- artificial precipitation 人工降水，人工降雨
- ice crystal 冰晶

译文：人工降水是指当云中降水条件不足，用飞机在云中飞行扰动和搅拌空气，并播撒催化剂，促进云中冰晶或云雾滴迅速形成，进而达到雨雪下降。

7. For the latter two manners, they are usually operated with large helicopters, while for the daily patrol, it is mostly performed with low-cost fixed-wing aircrafts, and recently the real-time image monitoring of UAVs is also alternatively applied for that end.

- fixed-wing aircraft 固定翼飞机
- real-time image monitoring 实时图像监控
- UAV 无人驾驶飞机简称"无人机"，英文缩写为"UAV"，是利用无线电遥控设备和自备的程序控制装置操纵的不载人飞机，或者由车载计算机完全地或间歇地自主地操作。目前在航拍、农业、植保、微型自拍、快递运输、灾难救援、观察野生动物、监控传染病、测绘、新闻报道、电力巡检、救灾、影视拍摄、制造浪漫等领域的应用，大大地拓展了无人机本身的用途，发达国家也在积极扩展行业应用与发展无人机技术。

译文：对于后两者，通常采用大型直升机飞行，而日常巡查，通常采用成本较低的固定翼飞机，近年来还出现了采用无人机实时图像监控的技术。

8. Aircrafts, if assembled with sprinkling system, can be used for pest control,

fertilization and weeding for forests, grasslands and crops, and in another way, if assembled with seeding system, can conduct aerosowing for forestation, grass planting and seeding of rice and soybean as well as protect forest from fires.

— assemble with　与……组装

译文：航空器装上喷撒（洒）系统，可进行森林、草地和农作物的病虫防治、施肥、除草。装上播撒系统，可飞播造林、植草、播种（如播水稻、大豆）等，还可进行护林防火。

9. After World War Ⅱ, a variety of insecticides, fungicides and herbicides sprang up in the market which necessitated the use of efficient spraying machines and tools to keep up with the work requirements. At the same time, large numbers of military aircrafts and pilots were transferred to agricultural sector, paving the way to the booming of agricultural and forestry aviation.

— spring up　出现；涌现；萌芽

— keep up with　赶得上；和……保持联系

— pave the way to　为……铺平道路；为……做准备

译文：第二次世界大战后，各种杀虫剂、杀菌剂和除草剂大量问世，要求用高效能的喷撒机具来满足工作的需要；与此同时，大量军用飞机和驾驶员转向农业，遂使农林业航空迅速发展。

10. The agricultural and forestry aviation is one of the most important parts of general aviation industry in China, but, as limited by the small scale and scattered operation of agricultural operators in China, it is in a sustained and stable but low growth.

— small scale　小规模

译文：农林航空是我国通用航空产业中最重要的组成部分。但是受到我国农业经营者规模小、经营分散的影响，农林航空作业的增长速度不高，处于一个持续稳定增长的状态。

11. GA-200 Fatman is a two-seater agricultural and forestry aircraft developed by Australia-based Gippsland Aeronautics Co., Ltd.. Founded in 1971, the company is mainly specialized in the modification business of aircrafts including wooden aircraft and turboprop aircraft, and it also develops GA-200 Fatman and GA-8 AirVan on its own.

— specialized in　专门从事，专门研究；专门经营

译文：GA-200"法特曼"是澳大利亚吉普斯兰（Gippsland）航空有限公司研制的双座农林飞机。该公司成立于1971年，主要从事包括木质飞机和涡桨飞机在内的飞机改型业务，自行研制的飞机包括GA-200"法特曼"和GA-8"空中大篷车"两种。

12. EMB-202 Ipanema is a small single-seater agricultural and forestry aircraft designed by Embraer S.A. and manufactured by Nevada Aviation Industry. Nevada Aviation Industry was founded in 1954 and then turned into a wholly-owned subsidiary of Embraer through acquisition on March 10, 1980.

— Embraer S.A.　巴西航空工业公司（Embraer S.A.），是巴西的一家航空工业集团，成立于1969年，业务范围主要包括商用飞机、公务飞机和军用飞机的设计制造，以及航空服务。现为全球最大的120座及以下商用喷气飞机制造商，占世界支线飞机市场约45%市场份额。

该公司现已跻身于世界四大民用飞机制造商之列，成为世界支线喷气客机的最大生产商。

- wholly-owned subsidiary 全资子公司，指的是完全由唯一一家母公司所拥有或控制的子公司。母公司可以通过两种方式来设立全资子公司：第一种是，从头开始成立一家新公司并修建全新的生产设备（例如工厂、办公室和机器设备等）；第二种是，收购一家现有的公司并将其设备纳为己用。

译文：EMB-202"伊帕内马"是由巴西航空工业公司设计、巴西内瓦航空工业公司制造的一种小型单座农林飞机。内瓦航空工业公司成立于1954年，1980年3月10日被巴西航空工业公司收购，成为其全资控股子公司。

## Exercises

Section one: Answer the following questions according to the text.

1. What is single engine aerial tanker (SEAT) and what is SEAT used for?
2. Why the manufacture and sales of worldwide agricultural and forestry aircrafts will embrace a steady and solid growth in the near future?
3. Why the agricultural and forestry aviation is in a sustained and stable but low growth in China?
4. Please briefly describe the development of agricultural and forestry aviation in some developed countries.
5. What is artificial precipitation? What are the main purposes of artificial precipitation?

Section two: Match the words or phrases in column A with the definition in column B.

| A | B |
| --- | --- |
| 1. artificial | a. any plant that crowds out cultivated plants |
| 2. rainfall | b. in agreement or consistent or reliable |
| 3. chemical | c. a small hard fruit |
| 4. manual | d. susceptible to attack |
| 5. spread | e. contrived by art rather than nature |
| 6. hazard | f. water falling in drops from vapor condensed in the atmosphere |
| 7. vulnerable | g. relating to or used in chemistry |
| 8. consistent | h. requiring human effort |
| 9. seed | i. process or result of distributing or extending over a wide expanse of space |
| 10. weed | j. a source of danger |

Section three: Translate the following into Chinese.

1. Agricultural aviation can greatly improve labor productivity and play an

important role in the prevention of natural disasters, pest control, improvement of human living environment and ecological balance.

2. The consistent high and stable yield of crops depends on the increasingly higher management skills, whereas currently the poor field management in the middle and later stages of crops stands as the main issue in China's agricultural production.

3. Machine works in the cropland could easily trip or overwhelm the plants and may even bruise the plants' root system and more importantly, it is difficult to realize a uniform and rapid control if there're vast areas of fields.

4. Pests like grass moth, East Asian migratory locust and army worm have all wreaked havoc due to the essential slowness of manual control.

5. Agrochemical dispersing operations by aircraft are efficient, fast and effective in disease and pest control. Also, as it applies one tenth as many amounts of chemicals as manual spraying at best, it is low in costs.

6. The sprinkled chemicals can also wipe out the moving pathogens and pests in the air, which will comprehensively hold down the infection of pathogens and keep control of its spread.

7. Apart from alleviating drought in the agricultural and pastoral areas, artificial precipitation can also reduce the fire risk and help extinguish fires in forest regions.

8. Nevertheless, in China, the forest coverage is relatively low; therefore the workloads of aerial forest fire-protection remain relatively alike in each year.

**Section four: Translate the following into English.**
1. 州长否认非法挪用州政府的资金。
2. 一些公司获得了这些产品的销售许可。
3. 这个国家最大的两家银行正计划合并。
4. 它被评为美国最适合居住的城市。
5. 新发现表明女性应当密切关注她们的胆固醇水平。
6. 航天器将绕着地球飞行以获取飞向木星所需的动力。
7. 当地市政会刚刚成立了一个研究资源再生利用问题的委员会。
8. 如果你要干教师这一行,必须要有充沛的精力。
9. 英语的一些特点是许多语言所共有的。
10. 与许多母亲一样,她觉得家庭与工作难以兼顾。

# 4.3 Air tourism
## 空中游览

## Background

Low-altitude airspace usually refers to the airspace for military aviation and civil aviation within an altitude of 1000 meters. And the low-altitude tourism is people's travel, entertainment and sport activities within the low-altitude airspace through general aviation transport, general aircraft and low-flying aircraft. Air tourism is the flight activities in which tourists take aircraft (airplane, helicopter, airship, ballon and paraglider) to engage in sightseeing, recreation and stunts experiencing in the airspace above certain designated areas, generally i.e. in some scenic spots and cities.

### International low-altitude tourism development

To protect the ecological environment, developed countries such as Europe, America and Australia choose not establishing ground transportation facilities in ecologically fragile areas but developing low-altitude tours instead, such as helicopter tour, which does not necessitate the prior establishment of runway, large parking apron, highways and railways, nor does cause land expropriation and destruction to vegetative cover on earth surface, and which on the other hand, is environmentally friendly with trivial impact on surrounding environment, and is convenient, fast, time-saving and less susceptible to the landform while tourists relocate to sightsee. Tourists can experience and watch attractions however distant and steep such attractions are, and to their satisfaction for the customized and high-end demands with on-spot involvement.

Low-altitude tourism has attracted the attention of the governments of developed countries. Many countries have formulated long-term development strategies and corresponding policies, invested funds, trained relevant professionals, and established a sound law, regulation and standard system as well as GA industry organizations, which altogether put the whole international low-altitude tourism on the track of fostering a variety of industrial businesses with the core theme of low-altitude tourism. For instance, Niagara Falls Park offers helicopter and hot-air balloon tours; every day in Colorado Canyon, dozens of helicopters take off one after another to render air cruise service to tourists; Helicopter sightseeing in Rotorua City of New Zealand tours across the Taravila Mountains to demonstrate the wonderful natural volcanic landscape of Rotorua; Tourists for Australian Great Barrier Reef, by the ride of helicopter, pass across the Coral Sea to dive, boat, and enjoy the wonderful experience; Visitors to Victoria Falls in Zambia take helicopter to witness the spectacle of Zambezi River rushing into the Bartoka Gorge, and to appreciate a top view of those elephants, hippos, rhinos, giraffes and other wildlife on the spot of your journey.

## Main reasons for the lagged air tourism in China

1. Low spending power. Nevertheless, with the growing economy and the rising demand contributed by foreign tourists and some domestic business travelers and middle-to-high-end tourists, such restraint is basically fading away.

2. Complicated approval formalities. Taking into consideration of the time and the space, the air tourism requires the plane to take off and land as closer as possible to scenic spots, whereas the fact is that the majority of civil aviation airport in China currently cannot meet such requirement, hence in order to expand air tourism business in China, establishing take-off and landing field (point) shall be put in the first place. However, regulations regarding the approval of such field (point) are extremely complicated, which involves military authority, civil aviation, local authorities and many other bodies and procedures. In addition, several regulations are not so clear-cut that many projects are aborted in the process of seeking for approval. Therefore, people liken the approval of taking-off and landing point as a walk through the maze. With a helpful guide, you may manage to walk it through while it will be fairly challenging if you feel the way on your own. Nevertheless, as national policies are right on the path of reform and improvement, such restraint is playing down.

3. Backward business model. Basically the typical business model of air tourism in China is to ensure not to make loss by renting aircrafts, which means that air tourism operator rent aircrafts from aircraft owners to operate their business. Such model theoretically can complement each party's advantages, whereas in actual practice, each collaboration party is driven to maximize his own interests in the shortest possible time, causing several situations unfavorable for the development of air tourism business. For instance, due to the high operation cost, the tourism ticket is highly priced and becomes market-unfriendly; and each party focuses on the short-term profits by furnishing primitive hardware facilities in the take-off and landing point as well as insufficient soft powers such as publicity and service, which altogether result in an awkward dilemma where the affordable middle-to-high-end tourists are afraid to be taken in and the desirable tourists cannot afford the ticket price. The aircraft lessees suffer from high pressure both economically and psychologically, and in most cases, such operators fail to continue their business for over one year either due to their economic strength or psychological endurance, many of whom fails in less than half an year. However, the successful experience of air tourism in other countries is that in the first half of the year when the new air tourism

project starts business, the main customers are walk-in tourists, and in the latter half of the year, there come several tour groups, and after one year, there come tour groups in bulk. To conclude, most of air tour operators in China have already given up before the advent of the business success.

### Low-altitude tourism development in China

China strongly encourages the development of tourism and proposes to actively develop low-altitude flight tourism. As an emerging tourism plan, low-altitude tourism is also listed as the key nationally supported tourism product. Low-altitude tourism harmonizes the "aviation" and the "tourism", and effectively fills the gap of tourism market in the transition of traditional scenic spots to all-for-one tourism.

Since the 12th Five-Year Plan, there are surging successful cases in the businesses of general aviation industry in China, including agricultural reclamation and spraying, power utilities inspection, offshore oil, flight training, and the emerging low-altitude tourism in recent years. In 2012, the low-altitude airspace in China has been expanded to the whole northeast and central-south regions, which help make way for commuter flying in low-altitude airspace. Low-altitude tourism, with its low demand for natural resources, caters to the requirement of harmonious development between tourism and ecological conservation, and plays a significant role in developing a sustainable tourism industry. In China, damages to the ecological environment take place again and again in the cause of thriving tourism, and in particular some areas suffer from irreversible ecological disasters. Therefore it's bound to see a new ecologically and environmentally friendly tourism industry burgeoning.

## General Vocabulary

airship  *n.* 飞艇
paraglider  *n.* 滑翔伞
stunts  *n.* 绝技（stunt 的复数）；惊人表演
scenic spot  风景区；景点
land expatriation  土地征收，土地征用
trivial  *adj.* 不重要的，琐碎的；琐细的
susceptible  *adj.* 易受影响的；易感动的；容许……的
relocate  *vt.* 重新安置；迁移
steep  *adj.* 陡峭的；不合理的；夸大的；急剧升降的
satisfaction  *n.* 满意，满足；赔偿；乐事；赎罪
customized  *adj.* 定制的；用户化的
on-spot  *adj.* 现场的；当场的
cruise  *v.* 乘船游览；以平稳的速度行驶；巡航，巡游，漫游；开车兜风；轻而易举赢得；猎艳（非正式）
Rotorua City  罗托鲁瓦市

Taravila Mountains　塔拉威拉山
Great Barrier Reef　澳大利亚的大堡礁
dive　*v.* 潜水，下潜；跳水；（飞机或鸟）俯冲；冲，奔，扑；突降，暴跌；迅速将手伸入；假摔
Victoria Falls　维多利亚瀑布
Zambia　*n.* 赞比亚（非洲中南部国）
witness　*vt.* 目击；证明；为……作证
spectacle　*n.* 景象；场面；奇观；壮观；盛大的演出；（复）眼镜
Zambezi River　赞比西河（非洲南部河流）
Bartoka Gorge　巴托卡峡
hippo　*n.* 河马
rhino　*n.* 犀牛（与rhinoceros同）
giraffe　*n.* 长颈鹿
wildlife　*n.* 野生动植物
lagged　延迟
middle-to-high-end　中高端
restraint　*n.* 抑制，克制；约束
fade　*vi.* 褪色；凋谢；逐渐消失
in the first place　首先，起初，第一
extremely　*adv.* 非常，极其；极端地
clear-cut　*adj.* 清晰的；轮廓鲜明的
abort　*vi.* 流产；堕胎；夭折；发育不全
challenging　*adj.* 挑战的；引起挑战性兴趣的
reform　*v.* 改革，革新；重组；（使）改过自新；（石油炼制）重整
backward　*adj.* 向后的；倒退的；（人）智力迟钝的；落后的；（场地位置）线后的
theoretically　*adv.* 理论地；理论上
collaboration　*n.* 合作；勾结；通敌
unfavorable　*adj.* 不宜的；令人不快的；不顺利的
hardware　*n.* 计算机硬件；五金器具
insufficient　*adj.* 不足的；不能胜任的，缺乏能力的
publicity　*n.* 宣传，宣扬；公开；广告；注意
awkward　*adj.* 尴尬的；笨拙的；棘手的；不合适的
dilemma　*n.* 困境；进退两难；两刀论法
affordable　*adj.* 负担得起的
desirable　*adj.* 可取的，值得拥有的，令人向往的；引起性欲的，性感的
psychologically　*adv.* 心理上地；心理学地
economic strength　经济实力；经济力量
walk-in　*adj.* 无预订散客，（宽敞得可以）步入的；（无需、未经）预约的
tour group　旅行团；旅游团队
in bulk　整批，散装；大批，大量

advent    n. 到来；出现；基督降临；基督降临节
emerging    adj. 走向成熟的；新兴的
harmonize    vt. 使和谐；使一致；以和声唱
all-for-one tourism    全域旅游
surging    v. 涌，涌动，汹涌；使强烈感到；（物价等）激增；（电流）浪涌（surge的现在分词）
reclamation    n. 开垦；收回；再利用；矫正
agricultural reclamation    农垦
irreversible    adj. 不可逆的；不能取消的；不能翻转的
bound to    必然；一定要

## Notes

1. Air tourism is the flight activities in which tourists take aircraft (airplane, helicopter, airship, ballon and paraglider) to engage in sightseeing, recreation and stunts experiencing in the airspace above certain designated areas, generally i.e. in some scenic spots and cities.
   - paraglider    n. 滑翔伞，滑翔伞（paraglider）飞行运动员驾翼型伞衣，利用空气升力起飞翱翔的一项航空运动。
   - engage in    从事于（参加）
   - scenic spot    风景区；景点

译文：空中游览是指游客搭乘航空器（飞机、直升机、飞艇、气球、滑翔伞）在特定地域上空进行观赏、游乐和特技体验的飞行活动。通常情况下，空中游览项目主要在一些风景名胜区及城市开展。

2. To protect the ecological environment, developed countries such as Europe, America and Australia choose not establishing ground transportation facilities in ecologically fragile areas but developing low-altitude tours instead, such as helicopter tour, which does not necessitate the prior establishment of runway, large parking apron, highways and railways, nor does cause land expropriation and destruction to vegetative cover on earth surface.
   - ecological environment    生态环境
   - ground transportation    地面运输
   - land expropriation    土地征收，土地征用

译文：欧美澳等发达国家为保护生态环境，放弃在生态脆弱地区建地面交通设施，转而发展低空旅游，如直升机旅游，直升机旅游不需建跑道和大型停机坪，也无需先行建成公路和铁路，不需征用土地，不破坏地表植被。

3. Tourists can experience and watch attractions however distant and steep such attractions are, and to their satisfaction for the customized and high-end demands with on-spot involvement.
   - high-end demand    高需求
   - on-spot    在现场，当场的

译文：旅游"触角和眼睛""无远不达、无险不及"，满足旅游的体验性、个性化和高层次需求。

4. Many countries have formulated long-term development strategies and corresponding policies, invested funds, trained relevant professionals, and established a sound law, regulation and standard system as well as GA industry organizations, which altogether put the whole international low-altitude tourism on the track of fostering a variety of industrial businesses with the core theme of low-altitude tourism.

译文：各国纷纷制订了长远发展战略、相应政策并投入资金；建立了完善的法规和标准体系并培育了相关专业人才，建立通用航空行业组织机构，使得整个国际低空旅游的发展，形成了众多以低空旅游为核心主题的产业门类。

5. Helicopter sightseeing in Rotorua City of New Zealand tours across the Taravila Mountains to demonstrate the wonderful natural volcanic landscape of Rotorua; Tourists for Australian Great Barrier Reef, by the ride of helicopter, pass across the Coral Sea to dive, boat, and enjoy the wonderful experience.

- volcanic landscape　火山景观
- Great Barrier Reef　大堡礁（英文：The Great Barrier Reef，法文：Grande barrière de corail），是世界最大最长的珊瑚礁群，位于南半球，它纵贯于澳大利亚的东北沿海，北从托雷斯海峡，南到南回归线以南，绵延伸展共有2011千米，最宽处161千米。有2900个大小珊瑚礁岛，自然景观非常特殊。大堡礁于1981年被列入世界自然遗产名录。

译文：新西兰罗托鲁瓦城的直升机观光飞越塔拉威拉山，展示罗托鲁瓦神奇的自然火山景色；澳大利亚大堡礁的游客乘坐直升机掠过珊瑚海去潜水、泛舟，享受奇妙感觉。

6. However, regulations regarding the approval of such field (point) are extremely complicated, which involves military authority, civil aviation, local authorities and many other bodies and procedures. In addition, several regulations are not so clear-cut that many projects are aborted in the process of seeking for approval.

- military authority　军事当局
- clear-cut　清楚的，明确的；轮廓鲜明的
- seek for　v. 寻找；追求；探索

译文：有关起降场（点）的报批规定极其复杂，涉及军方、民航、地方等多个部门和环节，而且部分规定并不透明，造成不少项目在报批阶段就已夭折。

7. Backward business model. Basically the typical business model of air tourism in China is to ensure not to make loss by renting aircrafts, which means that air tourism operator rent aircrafts from aircraft owners to operate their business.

译文：经营模式落后。我国的航空旅游业务基本采取中国特有的"保底租机"的经营模式，即航空旅游经营者向飞机拥有者租机开展经营。

8. For instance, due to the high operation cost, the tourism ticket is highly priced and becomes market-unfriendly; and each party focuses on the short-term profits by furnishing primitive hardware facilities in the take-off and landing point as well as insufficient soft powers such as publicity and service, which altogether result in an awkward dilemma where the affordable middle-to-high-end tourists are afraid to be

taken in and the desirable tourists cannot afford the ticket price.

— operation cost　运营成本

— result in　导致，结果是

译文：诸如：由于经营成本高，抬高了旅游票价，市场难以接受。注重短期利益，起降点各项硬件建设简陋，宣传和服务等软件建设严重不足，造成"消费得起的中高档游客不敢来，敢来的又嫌票价高"这样一种高不成低不就的局面。

9. However, the successful experience of air tourism in other countries is that in the first half of the year when the new air tourism project starts business, the main customers are walk-in tourists, and in the latter half of the year, there come several tour groups, and after one year, there come tour groups in bulk.

— tour group　旅行团；旅游团队

— in bulk　整批，散装；大批，大量

译文：而国外航空旅游的成功经验是，新的航空旅游项目开始的前半年，基本以散客为主，半年到一年，开始有旅游团队，一年以上就有大批量的旅游团队了。

10. Low-altitude tourism harmonizes the "aviation" and the "tourism", and effectively fills the gap of tourism market in the transition of traditional scenic spots to all-for-one tourism.

— fill gap　填补缺口，填补空白

— all-for-one tourism　全域旅游，是指在一定区域内，以旅游业为优势产业，通过对区域内经济社会资源尤其是旅游资源、相关产业、生态环境、公共服务、体制机制、政策法规、文明素质等进行全方位、系统化的优化提升，实现区域资源有机整合、产业融合发展、社会共建共享，以旅游业带动和促进经济社会协调发展的一种新的区域协调发展理念和模式。

译文：低空旅游将"航空"和"旅游"有机结合，在传统景区向全域旅游发展中，有效填补旅游市场的空白。

11. Low-altitude tourism, with its low demand for natural resources, caters to the requirement of harmonious development between tourism and ecological conservation, and plays a significant role in developing a sustainable tourism industry.

— cater to　迎合；为……服务

译文：低空旅游以其自然资源需用度小的优势，迎合了旅游与生态和谐发展的需求，对建设可持续发展的旅游业起到很大作用。

## Exercises

### Section one: Answer the following questions according to the text.

1. What are definitions of the low-altitude tourism and Air tourism?
2. Why do developed countries choose developing low-altitude tours instead of establishing ground transportation?
3. What are main reasons for the lagged air tourism in China?
4. What is the current situation of Low-altitude tourism development in China?
5. Pleas briefly describe the international low-altitude tourism development.

**Section two: Match the words or phrases in column A with the definition in column B.**

| A | B |
|---|---|
| 1. afford | a. come out into view, as from concealment |
| 2. entertainment | b. a demanding or stimulating situation |
| 3. sightsee | c. state of uncertainty or perplexity |
| 4. destruction | d. of concern to or concerning the internal affairs of a nation |
| 5. attraction | e. be able to spare or give up |
| 6. appreciate | f. an activity that is diverting and that holds the attention |
| 7. domestic | g. the termination of something by causing damage to it |
| 8. challenge | h. visit famous or interesting sights |
| 9. dilemma | i. recognize with gratitude; be grateful for |
| 10. emerge | j. the force by which one object attracts another |

**Section three: Translate the following into Chinese.**

1. In China, damages to the ecological environment take place again and again in the cause of thriving tourism, and in particular some areas suffer from irreversible ecological disasters.

2. Since the 12th Five-Year Plan, there are surging successful cases in the businesses of general aviation industry in China, including agricultural reclamation and spraying, power utilities inspection, offshore oil, flight training, and the emerging low-altitude tourism in recent years.

3. China strongly encourages the development of tourism and proposes to actively develop low-altitude flight tourism. As an emerging tourism plan, low-altitude tourism is also listed as the key nationally supported tourism product.

4. The aircraft lessees suffer from high pressure both economically and psychologically, and in most cases, such operators fail to continue their business for over one year either due to their economic strength or psychological endurance, many of whom fails in less than half an year.

5. In addition, several regulations are not so clear-cut that many projects are aborted in the process of seeking for approval. Therefore, people liken the approval of taking-off and landing point as a walk through the maze.

6. With the growing economy and the rising demand contributed by foreign tourists and some domestic business travelers and middle-to-high-end tourists, such restraint is basically fading away.

7. Helicopter tour is environmentally friendly with trivial impact on surrounding environment, and is convenient, fast, time-saving and less susceptible to the landform while tourists relocate to sightsee.

**Section four: Translate the following into English.**
1. 这名才华横溢的年轻小提琴手曾在世界各地的音乐会舞台上演出。
2. 美国的人口死亡率在20世纪60年代进入平稳阶段后突然下降。
3. 我通过节食体重每周减轻1磅左右,但最近体重稳定了下来,1磅也没少。
4. 需要有一位真正具有个人魅力的领导人来振兴该党。
5. 法国大革命使法国由君主政体变成共和政体。
6. 没有任何计算机比得上人脑复杂。
7. 很多人想要得到这份工作,竞争非常激烈。
8. 在特别易感的儿童中,这种疾病病情发展会非常迅速。
9. 争论是源于对法律的两种截然不同的解释。
10. 幸好目前公司的银行账户里还有结余。

# 4.4 Industrial aviation
## 工业航空

## Background

### 1. Introductions to industrial aviation

(1) Concept of industrial aviation

Industrial aviation, an important constituent part of GA flight operations, is a general term for all kinds of flight activities associated with industrial production through the use of aircrafts, such as aerial photography, aerogeophysical prospecting, aviation petroleum, aerial remote sensing, aerial patrol and aerial hanging. Industrial aviation is operated high above the ground and is capable of conducting various production activities that human beings are unable to implement on the ground, so it plays an important role in industrial production and construction.

(2) Functions and advantages of industrial aviation

The primary task of industrial aviation flight is not to transport passengers or goods as regular airlines from one point to another point, but to perform aerial activities related to industrial production for the following functions.

① For detection, including aerial mineral prospecting, aerial remote sensing etc;

② For patrol, including aerial patrol, environmental monitoring, marine monitoring etc;

③ For service, including assistance to urban construction, water diversion, aviation petroleum, advertisement etc;

④ For production, including aerial hoisting, aerial mapping, live working etc.

The advantages of performing industrial production activities with aircrafts include the following:

① Working at heights can accomplish those production activities that ground-based mankind can't perform, such as aerial video-shooting and photo-taking, etc.

② It can cut down production cost and improve work efficiency. For example, considering the fast and maneuverable nature of aircrafts, mineral prospecting and remote sensing by aircrafts could be dozens or hundreds time as efficient and accurate as that done at level ground.

③ It can enhance production safety. For example, general aviation, if applied in the risky offshore petroleum exploration, can effectively improve safety level by providing fast and convenient transport, emergency rescue and other safe services.

## 2. Typical flight operations of industrial aviation

Industrial aviation includes a variety of flight operations, and due to the space limitations, we will touch on the following several typical industrial aviation operations only.

### (1) Aerial photography

Aerial photography is such a GA operation that takes aerial photograph of ground sceneries with photographic equipment mounted on the GA aircrafts such as airplane, balloon, etc. It is not only intensively applied for mapping and surveying, but also in many other fields like national economic construction, military and scientific research.

Aerial photography can be dated back to 1950s. By that time, cameras were carried in balloons to take photos of cities for casual enjoyment only, but it opened up the history of observing the earth from the air. In 1909, W. Wright of the United States photographed the ground from an airplane for the very first time. Since then, in pace with the rapid technical advancement of aircrafts and flight as well as cameras and photographic materials, the aerial photos have been of increasingly higher quality and wider application.

Based on how the photograph is taken, aerial photography can be classified into single image photography, strip photography and area photography. Single image photography is to take photograph of a single fixed object and in general one photo or a pair of photos is taken. Strip photography is to take successive photographs of a long narrow area or linear thing (railway, high way, etc.) along one flight course. Area photography (or known as regional photography) means successive photography for the large area along multiple flight courses.

### (2) Aerial geophysical prospecting and remote sensing

Aerial geophysical prospecting and remote sensing is an application of general aviation in which on the platform of aircrafts, geophysical or chemical methods (such as remote sensing, electromagnetic, energy spectrum, light spectrum and gravity) are used to investigate and monitor basic geology, geological resources, ecological environment, etc. The typical operations of aerial geophysical prospecting and remote sensing include ① basic geological survey; ② mineral resources exploration; ③ survey and monitoring of geological disaster; ④ survey and dynamic monitoring of land resources and ecological environment; ⑤ hydrogeological and engineering geological survey and monitoring; ⑥ survey and dynamic monitoring of marine and coastal resources; ⑦ urban resources survey and digital city, etc. And as to those aircrafts, the platform of aerial geophysical prospecting and remote sensing, the majority of them are small paddle-shaped fixed-wing aircrafts including Y-12, Y-11, Y-8, Y-5, Cessna 208, DHC-6-300 Twin Otter, and Antonov An-30 as well as helicopters such as AS350B2, Mi-8, etc.

Aerogeophysical prospecting is particularly fitful for exploration and survey work in complex areas. Important mining areas and their peripheral areas are mostly characterized with complex landforms, which will lower working efficiency of geophysical exploration at ground level and pose great challenge on construction work as well. Furthermore, given that most of workspace for ground-based geophysical survey covers relatively small areas, the depth of geophysical exploration and the accuracy of peripheral exploration could be compromised to certain extent. However, if taking the aircraft and helicopter as the platform, we can make full use of their flight performance to survey and measure along the undulating terrain, which has the advantages of high speed, large measurement area and complete information.

In the operation of aerogeophysical prospecting, helicopters outshine fixed-wing airplanes in terms of measurement accuracy and anomaly resolution. The geophysical survey system of helicopter can be mounted as pod type or the type of rigid frame. Among them, the pod-typed survey system is generally composed of electromagnetic system, magnetometer, data collection system, GPS navigation and positioning system, height measurement system, analog recorder, power system, etc. And the rigidly framed survey system consists of aerial magnetometer, traveling magnetic compensator, GPS navigation and positioning equipment, GPS differential positioning equipment, 7K data display equipment, etc.

(3) Offshore petroleum service

Offshore petroleum service, also simply referred to as oversea flying, is air transportation by helicopters among offshore oil drilling platform, oil extraction platform, platform supply vessel and land area. Its main tasks include the transportation of commuters and much-needed equipment, instrument and geological data, first-aid treatment of the injured or the sick, emergency evacuation of personnel prior to the strike of typhoon, search and rescue post maritime distress, aerial firefighting etc. The oversea flying has the following characteristics: its flight zone is way far away from the coast; it is not easy to be alerted earlier to the weather change as there is a lack of sufficient navigation equipment and meteorological data; the landing platform has a small area and is also close to different obstacles which create certain difficulties for navigation, take-off and landing.

To assure the safety of oversea flying, there're some special requirements for offshore petroleum service. For example, oversea flying must be equipped with pontoons, sufficient fuels and protective materials that are accessible to flight crew and staff like lifejackets, lifeboats, pharmaceuticals, shark repellents, crash radio, signal guns. And the flight crew must meet the specified flight standards, with competent flight experience over the sea and a minimum flight time of 20 hours. Moreover, weather changes must be forewarned during oversea flight and the flight can be given green light only if necessary flight standards are followed.

(4) Aerial hanging

Aerial hanging, also known as hanging operation refers to the general aviation operation of lifting and installing the ground-based materials and equipment by taking advantage of the helicopter's carrying capacity. Aerial hanging is applied mainly for two purposes. One is to lift to transport, meaning that materials on the ground are transported to a place by lifting, like the wood felled from the forest are lifted and then in that state transported to the specified place. The other is to lift to install, which means the equipment on the ground are installed by means of helicopter's loading capacity, such as the installation of power transmission equipment and aerial stringing, etc.

Aerial hanging can be executed exclusively by helicopter because helicopter can hover in the air whereas fixed-wing aircraft cannot. Aerial hanging is able to fulfill the transportation work without ground level road and the installation work when the lifting equipment cannot be installed in the workplace. As the special application of helicopters, aerial hanging must take into account of the lifting capacity of the helicopter to guarantee a safe flight. Helicopters deployed for hanging operations usually features a relatively greater loading capacity, like the total lifting capacity of Russia-made Mi-26 is as great as 26t, and it is thus widely used in the construction of hydropower stations in remote areas, the lifting and installation of power transmission lines as well as the logging and transportation of forest timber.

### General Vocabulary

constituent   *adj.* 构成的；选举的
operation    *n.* 运营；运作；业务操作
associate    *adj.* 关联的；联合的
photography  *n.* 摄影；摄影术
prospect     *vt.* 勘探，勘察

industrial  *adj.* 工业的；产业的；工业造型
construction  *n.* 建设；建造
advantage  *n.* 优势；利益；有利条件 *vt.* 有利于；使处于优势；*vi.* 获利
function  *vi.* 运行；活动；行使职责；*n.* 功能；
primary  *adj.* 主要的；初级的；基本的
regular  *adv.* 定期地；经常地
mineral  *adj.* 矿物的；矿质的
environmental  *adj.* 环境的，周围的；有关环境的
monitor  *v.* 监视，监听，监督
assistance  *n.* 援助，帮助；辅助设备
hoist  *v.* 升高；举起
accomplish  *vt.* 完成；实现；达到
maneuverable  *adj.* 有机动性的；容易操作的；可调动的
accurate  *adj.* 精确的
typical  *adj.* 典型的；特有的；象征性的
limitation  *n.* 局限性；（限制）因素；边界
scenery  *n.* 风景；景色；舞台布景
balloon  *n.* 气球
intensively  *adv.* 强烈地；集中地
economic  *adj.* 经济的，经济上的；经济学的
scientific  *adj.* 科学的，系统的
observe  *v.* 观察；遵守
successive  *adj.* 连续的；继承的；依次的；接替的
regional  *adj.* 地区的；局部的；整个地区的
multiple  *adj.* 多重的；多样的；许多的
remote  *adj.* 遥远的；偏僻的；疏远的
method  *n.* 方法，方式；研究方法
environment  *n.* 环境，外界
dynamic  *adj.* 动态的；动力的；动力学的；有活力的
hydrogeological  *n.* 水文地质；水文地质学
investigate  *v.* 调查；研究
spectrum  *n.* 光谱；频谱；范围；余象
fitful  *adj.* 一阵阵的；断断续续的；不规则的；间歇的
peripheral  *adj.* 外围的；次要的；（神经）末梢区域的
characterized  *adj.* 以……为特点的
furthermore  *adv.* 此外；而且
relatively  *adv.* 相当地；相对地，比较地
depth  *n.* 深度；深奥
compromise  *vt.* 妥协；连累
certain  *adj.* 某一；必然的；确信；无疑的；有把握的

measure　*vi.* 测量；估量
undulate　*v.* 使波动；使成波浪形
terrain　*n.* 地形，地势；领域；地带
outshine　*vt.* 使相形见绌；胜过；比……更亮；*vi.* 放光
anomaly　*n.* 异常；不规则；反常事物
electromagnetic　*adj.* 电磁的
position　*v.* 定位；放置
analog　*adj.* 模拟的
rigidly　*adv.* 严格地；坚硬地；严厉地；牢牢地
frame　*v.* 制订；建造
compensator　*n.* 补偿器；自耦变压器；赔偿者；补偿物
navigation　*n.* 航行；航海
differential　*adj.* 微分的；差别的；特异的
among　*prep.* 之中；跻身；当中
extraction　*n.* 萃取；家世；提取
include　*vt.* 包含，包括
emergency　*adj.* 紧急的；备用的
evacuation　*n.* 疏散；撤离；排泄
typhoon　*n.* 台风
maritime　*adj.* 海事的；海运；生在沿海
obstacle　*n.* 障碍；障碍物；阻碍
create　*vt.* 创造，创作；造成
navigation　*n.* 航行；航海
assure　*vt.* 保证；担保；使确信；弄清楚
requirement　*n.* 需求；要求；职位要求
sufficient　*adj.* 足够的；充分的
fuel　*n.* 燃料，供给燃料
protective　*adj.* 防护的；关切保护的；保护贸易的
pharmaceutica　*n.* 药物
repellent　*n.* 驱虫剂
specify　*v.* 指定；详细说明
hang　*v.* 悬挂
install　*v.* 安装
material　*n.* 材料；材质
string　*v.* 用带系上
exclusively　*adv.* 仅仅；专有地；唯一地
deploy　*v.* 部署；加派；展开；展开后
timber　*n.* 木材；木料

# Notes

1. Industrial aviation, an important constituent part of GA flight operations, is a general term for all kinds of flight activities associated with industrial production through the use of aircrafts, such as aerial photography, aerogeophysical prospecting, aviation petroleum, aerial remote sensing, aerial patrol and aerial hanging.

— associate with  联合；与……联系在一起；和……来往

— aerial photography  航空摄影

— aerogeophysical prospecting  航空物探航空地质调查方法（aerial geological survey），简称航空物探，是物探方法的一种。它是通过飞机上装备的专用物探仪器在航行过程中探测各种地球物理场的变化，研究和寻找地下地质构造和矿产的一种物探方法。目前已经应用的航空物探方法有航空磁测、航空放射性测量、航空电磁测量(航空电法)等。

译文：工业航空是使用航空器进行与工业生产相关的，如航空摄影、航空物探、航空石油服务、航空遥感、航空巡线、空中吊装等各种飞行活动的总称，它是通用航空飞行作业的重要组成部分。

2. For detection, including aerial mineral prospecting, aerial remote sensing etc; For patrol, including aerial patrol, environmental monitoring, marine monitoring, etc;

— aerial mineral prospecting  航空探矿法是在飞机上进行物理探矿的方法，包括航空磁法、航空电磁法、航空放射性测量和航空红外测量等。

— aerial remote sensing  航空遥感又称机载遥感，是指利用各种飞机、飞艇、气球等作为传感器运载工具在空中进行的遥感技术，是由航空摄影侦察发展而来的一种多功能综合性探测技术。

译文：探测功能，具体包括航空探矿、航空遥感等。巡视功能，具体包括航空巡线、环境监测、海洋监测等。

3. For service, including assistance to urban construction, water diversion, aviation petroleum, advertisement, etc; For production, including aerial hoisting, aerial mapping, live working, etc.

— aviation petroleum  石油航空系以直升机为主工具，辅以小型固定翼飞机，为海洋和陆地石油、天然气资源勘探、开发和管理提供空中运输的作业。国际上亦称石油开发后勤支援。石油航空是随石油能源勘探、开发的发展而产生的。它具有迅速、灵活、机动的优点。它已成为石油勘探、开发重要的后勤保证手段。中国在塔里木盆地和沿海海域的石油勘探开发均利用了石油航空。

— aerial mapping  航空测绘是一种以大气层内的飞行器为测量载体的对地测绘手段，其测绘对象是地面物体的位置关系，目的是通过航空拍摄获得的数据来绘制大地坐标，其通常采用的方法是航空摄影测量。航空摄影测量是在飞机上利用航摄仪器对地面进行连续拍摄，绘制地形图的过程。

译文：服务功能，具体包括为城市建设提供帮助、引水作业、石油航空、广告宣传等。生产功能，具体包括空中吊装、航空测绘、带电作业等。

4. It can cut down production cost and improve work efficiency. For example, considering the fast and maneuverable nature of aircrafts, mineral prospecting and

remote sensing by aircrafts could be dozens or hundreds time as efficient and accurate as that done at level ground.

— cut down　削减；砍倒；杀死；删节；胜过

译文：能够降低生产成本，提高工作效率。例如，利用航空器快捷、机动性高的特点从事航空探矿、航空遥感，其工作的效率和准确性可以是在地面进行这类工作的几十倍甚至上百倍。

5. Aerial photography can be dated back to 1950s. By that time, cameras were carried in balloons to take photos of cities for casual enjoyment only, but it opened up the history of observing the earth from the air.

— date back to　追溯到；从……开始有

— open up　打开，开设，开辟，开创

译文：航空摄影始于19世纪50年代，当时从气球上用摄影机拍摄的城市照片，虽只有观赏价值，却开创了从空中观察地球的历史。

6. Strip photography is to take successive photographs of a long narrow area or linear thing (railway, high way, etc.) along one flight course. Area photography (or known as regional photography) means successive photography for the large area along multiple flight courses.

— Strip photography　航线航空摄影又称"带区航空摄影"，是沿宽度小而长度大的狭长地带所进行的航空摄影。通常沿一条或两条航线对地面狭长地区或线状地物进行连续摄影。航向相邻相片之间有一定重叠，一般为60%，不少于53%。常用于河流或道路的勘测。

— Area photography　沿数条航线对较大区域进行连续摄影，称为面积摄影（或区域摄影）。面积摄影要求各航线互相平行，在同一条航线上相邻相片间的航向重叠为60%～53%，相邻航线间的相片也要有一定的重叠，这种重叠称为旁向重叠，一般应为30%～15%。实施面积摄影时，通常要求航线与纬线平行，即按东西方向飞行，但有时也按照设计航线飞行。由于在飞行中难免出现一定的偏差，故需要限制航线长度一般为60～120千米，以保证不偏航，而产生漏摄。

译文：航线摄影是指沿一条航线，对地面狭长地区或沿线状地物（铁路、公路等）进行的连续摄影；面积摄影（或区域摄影）是指沿数条航线对较大区域进行连续摄影。

7. And as to those aircrafts, the platform of aerial geophysical prospecting and remote sensing, the majority of them are small paddle-shaped fixed-wing aircrafts including Y-12, Y-11, Y-8, Y-5, Cessna 208, DHC-6-300 Twin Otter, and Antonov An-30 as well as helicopters such as AS350B2, Mi-8, etc.

— Cessna 208　赛斯纳208是美国赛斯纳飞机公司研制的单发涡桨式多用途轻型飞机，除可用作客、货运输外，换装专用设备后，还可用于空中灭火、空中摄影、农业喷洒、边境巡逻、跳伞、空投物资、医疗救护和监视飞行等任务。塞斯纳208经历了一系列的修改，并衍生出不同的机型，由最初的型号演变出多种改型。塞斯纳208以其优良的适应能力著称，公司提供了不同的起落架安装模式，使塞斯纳208能适应不同的地形，甚至包括水上版本。

— DHC-6-300（双水獭）　是加拿大德·哈维兰飞机公司(现属加拿大庞巴迪公司)研制的20座双发涡轮螺旋桨式多用途短距起落运输机。

— Antonov An-30　安-30是从安-24系列发展而来的空中对地测绘飞机，它重新设计了一个新的机身框架，在机头加装透明观测窗，并将驾驶舱地板提高，使领航员获得了更好的

视界，且能够更方便的进出观察舱。新的机头里边装配了精确的导航设备，包括光学瞄准具，以确保航空摄影的准确性。

译文：航空物探遥感所用的航空器平台主要为小型桨状固定翼飞机和直升机，如运系列（Y-12、Y-11、Y-8、Y-5）、Cessna208（塞斯纳）、DHC-6-300（双水獭）、安-30等固定翼飞机和AS350B2、米-8等型号直升机。

8. However, if taking the aircraft and helicopter as the platform, we can make full use of their flight performance to survey and measure along the undulating terrain, which has the advantages of high speed, large measurement area and complete information.

- make full use of    充分利用，充足应用，充分使用
- undulating terrain    丘陵地带

译文：而以飞机和直升机为平台，则可充分利用它们的飞行性能，沿地形起伏飞行进行勘查和测量工作，具有速度快、测量面积大、信息完整丰富等优点。

9. Among them, the pod-typed survey system is generally composed of electromagnetic system, magnetometer, data collection system, GPS navigation and positioning system, height measurement system, analog recorder, power system, etc.

- composed of    合成，由……组成
- magnetometer    磁力仪是测量磁场强度和方向的仪器的统称。测量地磁场强度的磁力仪可分为绝对磁力仪和相对磁力仪两类。主要用途是进行磁异常数据采集以及测定岩石磁参数，从20世纪至今，磁力仪经历了从简单到复杂，机械原理到现代电子技术的发展过程。
- GPS navigation    全球定位系统导航（Global Position System），利用GPS定位卫星，在全球范围内实时进行定位、导航的系统，GPS是由美国国防部研制建立的一种具有全方位、全天候、全时段、高精度的卫星导航系统，能为全球用户提供低成本、高精度的三维位置、速度和精确定时等导航信息，是卫星通信技术在导航领域的应用典范，它极大地提高了地球社会的信息化水平，有力地推动了数字经济的发展。

译文：其中，吊舱式直升机测量系统一般由电磁系统、磁力仪、数据收录系统、GPS导航定位系统、高度测量系统、模拟记录仪、电源系统等组成。

10. Offshore petroleum service, also simply referred to as oversea flying, is air transportation by helicopters among offshore oil drilling platform, oil extraction platform, platform supply vessel and land area.

- oil drilling    石油钻井，探油，石油钻探
- oil extraction    油萃取；原油开采；石油抽提

译文：海上石油服务简称为海上飞行，是指使用直升机担负海上石油钻井平台、采油平台、后勤供应船平台与陆地之间的运输飞行。

11. Its main tasks include the transportation of commuters and much-needed equipment, instrument and geological data, first-aid treatment of the injured or the sick, emergency evacuation of personnel prior to the strike of typhoon, search and rescue post maritime distress, aerial firefighting etc.

- commuter    *n.*（远距离）上下班往返的人；
- emergency evacuation    应急疏散

— prior to 在……之前；居先

— typhoon  n. 台风，属于热带气旋的一种。热带气旋是发生在热带或副热带洋面上的低压涡旋，是一种强大而深厚的"热带天气系统"。台风灾害主要是在台风登陆前和登陆之后引起的。台风引起的直接灾害通常由狂风、暴雨、风暴潮三方面造成。

译文：其主要任务包括运送上下班的职工、急救伤病员、运输急需的器材、设备及地质资料、在台风前运送人员紧急撤离、发生海难事故后进行搜索与援救、空中消防灭火等。

12. For example, oversea flying must be equipped with pontoons, sufficient fuels and protective materials that are accessible to flight crew and staff like lifejackets, lifeboats, pharmaceuticals, shark repellents, crash radio, signal guns.

— equipped with  装备着……；有……的配置

— accessible to  易接近的；可归属的；可得到的

— crash radio  应急广播是指当发生重大自然灾害、突发事件、公共卫生与社会安全等突发公共危机时，造成或者可能造成重大人员伤亡、财产损失、生态环境破坏与严重社会危害，危及公共安全时，应急广播可提供一种迅速快捷的讯息传输通道，在第一时间把灾害消息或灾害可能造成的危害传递到民众手中，让人民群众在第一时间知道发生了什么事情，应该怎么撤离、避险，将生命财产损失降到最低。

译文：例如：海上飞行，必须安装浮筒，飞机上必须携带可供飞行人员、工作人员使用的救生衣、救生艇、药剂、驱鲨剂、应急电台、信号枪等防备物质。

13. Aerial hanging can be executed exclusively by helicopter because helicopter can hover in the air whereas fixed-wing aircraft cannot. Aerial hanging is able to fulfill the transportation work without ground level road and the installation work when the lifting equipment cannot be installed in the workplace.

— hover  vi. 盘旋，悬停是指航空器在一定高度上保持空间位置基本不变的飞行状态。在中国台湾，此术语也常称为"停悬"。具有这种飞行能力的航空器主要是直升机，这是它与固定翼航空器之间的一个重要差别。

译文：航空吊挂只能通过直升机进行，因为直升机可以在空中悬停，而固定翼飞机做不到。航空吊挂能够完成没有地面道路的运输，完成工作场地无法安装吊运设备不能进行安装而进行的安装作业。

## Exercises

Section one: Answer the following questions according to the text.

1. What is aerial hanging and what is it used for?
2. What are main tasks of offshore petroleum service?
3. Why helicopters are more favorable than fixed-wing aircrafts when performing aerial hanging services?
4. What do the typical operations of aerial geophysical prospecting and remote sensing include?
5. What is aerial photography? What services can aerial photography provide?

Section two: Match the words or phrases in column A with the definition in column B.

| A | B |
|---|---|
| 1. scientific | a. what something is used for |
| 2. industrial | b. a public promotion of some product or service |
| 3. typical | c. apart in space |
| 4. limitation | d. a general conscious awareness |
| 5. mineral | e. of or relating to the practice of science |
| 6. sense | f. the possibility of future success |
| 7. remote | g. of a feature that helps to distinguish a person or thing |
| 8. advertisement | h. a principle that limits the extent of something |
| 9. prospect | i. get from the earth by excavation |
| 10. function | j. of or relating to or resulting from industry |

Section three: Translate the following into Chinese.

1. Industrial aviation is operated high above the ground and is capable of conducting various production activities that human beings are unable to implement on the ground, so it plays an important role in industrial production and construction.

2. The primary task of industrial aviation flight is not to transport passengers or goods as regular airlines from one point to another point.

3. Working at heights can accomplish those production activities that ground-based mankind can't perform, such as aerial video-shooting and photo-taking, etc.

4. For example, general aviation, if applied in the risky offshore petroleum exploration, can effectively improve safety level by providing fast and convenient transport, emergency rescue and other safe services.

5. It is not only intensively applied for mapping and surveying, but also in many other fields like national economic construction, military and scientific research.

6. In pace with the rapid technical advancement of aircrafts and flight as well as cameras and photographic materials, the aerial photos have been of increasingly higher quality and wider application.

7. Based on how the photograph is taken, aerial photography can be classified into single image photography, strip photography and area photography.

8. Aerogeophysical prospecting is particularly fitful for exploration and survey work in complex areas.

Section four: Translate the following into English.

1. 他以0.01秒的优势打破了100米短跑的前世界纪录。
2. 让我非常惊讶的是，他们答应了我们所有的要求。
3. 每个人衡量成功的标准不尽相同——有些人认为快乐比金钱更重要。
4. 大多数人认同人们需要遵守一定的（行为）准则。

5. 虽然她的著作成书于20多年前，但时至今日仍被奉为考古学的权威读本。
6. 因为天气恶劣，渡船已经停开了。
7. 建筑工人站在用绳子从房顶上吊下来的木制平台上作业。
8. 当然，挑起事端的一方毫无疑问应对目前的冲突承担首要责任。
9. 购买全部6只平底锅只需半价——不要错过这难得的机会。
10. 我们在旅途中遇到了各种各样的问题，包括轮胎被扎破了。

# 4.5 Pilot license
## 飞行执照

## Background

The development of general aviation can directly and indirectly spur the burgeoning of many industries which then in return furthers GA's growth. And flight training is one of these industries.

### 1. Basic Concept

Pilot license training usually refers to the training of private or commercial pilot license. According to the interpretation of Provisions on the Administration of Business Licensing for General Aviation under CAAC's regulation No.CCAR-135, training for aircraft pilot license refers to the flight activities carried out with aircrafts for the purpose of mastering piloting skills and obtaining pilot license, which include any flight for normal teaching purpose, flight in the company of trainer and trainee's solo under the guidance of the trainer, but exclude skilled flight. In the training, qualified training schools will organize trainees to complete various training courses and subjects according to the requirements of flight training management of CAAC so as to make trainees meet the requirements of pilot's flight standards. And pilot license with different qualifications shall be obtained through the examination organized by CAAC.

### 2. Flight training model abroad

Flight training model can exert significant influence on the professional growth of pilots and shall be a key issue to be understood and researched in the flight training industry. Due to different national conditions, the mainstream model of pilot training in the powerful countries of civil aviation around the world varies widely with ours, and their training model has already been fully legalized and market-driven.

(1) Standardization of pilot training via legislation

In the United States, the conditions of establishing elementary flight training institutions, as well as the qualification standards of various pilots, and airline's recruitment conditions for pilots are all based on the relevant provisions of the Federal Aviation Regulations (FAR) issued by the Federal Aviation Administration (FAA) of the Department of Transportation. In the FAR, Article 61 to 67 under the subchapter D prescribes respectively the regulations on the qualification of pilots, flight instructors and other aviation personnel, and Article 141 under the subchapter H specifies legislative norms for the qualification of running pilot schools, requirement of personnel and facilities, training programs and curriculums, examinations, etc. And in other countries, similar regulations are also instituted. All in all, the oversea civil aviation flight schools have low conditions and can easily obtain the qualification certificate for civil aviation flight training, which then bring about a relatively large number of flight training organizations widely located in countries of strong civil aviation industry like the United States.

(2) Growth pattern of pilot

① Commercial pilot must possess adequate flight experiences

In the United States, it doesn't really take much to obtain the ordinary pilot qualification, yet with that, pilots still cannot directly join in airline transportation in airline companies. Pursuant to the provisions of American laws, pilot can be eligible to apply for pilot openings in transportation airlines only if he has performed accumulatively 1500-hour flight with small aircrafts as well as flight of certain hours with double-engined aircrafts and middle-sized aircrafts. Most of people gain those flight experiences by taking on a teaching position in aviation clubs or flight training organizations. Therefore, pilots undertaking teaching works in the aviation clubs or training organizations in the United States generally won't stay for long on teachings.

② Airlines employ pilots on contract basis

Upon obtaining all the qualifications of an airline pilot, you can't naturally become one of them, because it is only possible, provided that the opening for pilots arises in the airlines and meanwhile you pass all the relevant assessments in the recruitment. With the specified long flying experiences and so many candidate pilots, the rejections are understandably high in the process of airline pilot recruitment. Hence those who finally make it to be an airline transport pilot have comparatively more outstanding piloting skills as well as cultural and academic competence.

③ Model of basic pilot training

Pilots are mostly trained to pilot small aircrafts, and accordingly curriculums are devised and basic professional literacy is developed in tandem with that aim.

The vast majority of basic pilot training is not for academic education, but for vocational training. As a matter of fact, pilot training has no substantial difference from driving training in which trainees study traffic laws and basic driving skills of motor vehicle at vehicle driving school so as to get the driving license. Hence, in other countries, especially the United States, the curriculum design in flight training organizations is distinctly oriented for the purpose of obtaining a pilot license. All the training programs are closely focused on completing the required theoretical and practical learning in the shortest possible time and obtaining the corresponding flight experience so as to meet the requirements for the issuance of pilot license.

The flight training organizations are managed by category in accordance with the regulations. Subject to the internally established quality management system, the flight training organizations can be classified into flight schools under CCAR-141 and training institutions under CCAR-61. It is to be noted that such classification is made irrespective of the training objectives of pilots, which in other words, means that trainees, graduating from flight schools or training institutions, can only fly small aircrafts upon their completion of pilot training and also trainees from the fight schools are not necessarily eligible to directly work in airlines and pilot large passenger planes. Their difference lies in that those flight schools organized under CCAR-141build a more complete and sound internal quality management system and they can provide continuous trainings from level to level. Hence, as provided by the regulations, the flight time during the training of such flight schools under CCAR-141 can be less than that of training institutions under CCAR-61, which similarly brings about a lower training cost as well.

With regard to the implementation of flight training, basically it is taught and practiced by aviation theory session and flying skill training session for all countries around the world as all countries are identically bound by the relevant treaties and annexes of International Civil Aviation Organization (ICAO). In the training of flying skill, the training models in different foreign countries are roughly the same, which all follow the stage of PPL, CPL and IR. However, in the teaching of aviation theory, the United States and Canada adopt an integrated curriculum model in which ground teaching programs and theoretical examinations are conducted by the targeted category and level of certificates, namely PPL, CPL or IR etc. Differently, European countries, Australia, New Zealand and some other countries use sub-course model, meaning that based on the relevancy of theoretical knowledge and contents of aviation, teachings and examinations are done respectively by each individual course, such as flight principle, aircraft structure, propeller engine and jet engine, aviation meteorology, air navigation, instrument and air traffic control, and only after all courses are passed, can the trainee be recognized with qualified standard on aviation knowledge for the relevant pilot license and level.

## 3. Pilot training and its market development in China

### (1) The demand for pilots in the development of air transport

With the economic growth and the changes of pilot training system in China, private pilot training will emerge as a promising market. According to ICAO's statistics, globally, operational flights account for approximately 20% of the total flights, whereas the rest two types of GA flight, i.e. business and recreational flights in total take up more than 50% of that in which China barely contributes any. Nevertheless, as the popularization of recreational flights, more and more people will wish to obtain a private pilot license.

Also, considering the expansion of domestic civil aviation industry, the aircraft procurement will see increasing growth and the lack of pilots will become more intense. Enterprises need to invest a great deal of money in training pilots first and therefore most of them will attach stings of lifelong service to pilots, which will restricts the free flow of pilots among different companies in China. It is understood that in China it takes minimally about one million yuan to train a pilot before he gets to pilot and more than six million up to eight million yuan will be spent in training one single captain. Moreover, pilots should receive a variety of regular trainings on yearly basis, shortly after they work on their posts. Thus, pilots are usually deemed as private assets by airlines and will not be easily transferred out.

### (2) Current situation of pilot training in China

At the moment, pilots in Chinese airlines are mainly derived from flight cadets who graduate from the civil aviation flight schools in China and hold college diplomas as well as the professional pilot licenses before they are employed. In addition, pilots retired from air force and flight cadets independently trained by airlines with foreign pilot training institutions are also sources of airline pilots.

Apart from Civil Aviation Flight University of China who has accumulated pilot education experiences for 50 years and are complete with both the education and training conditions for the cultivation of pilots throughout the junior college stage, undergraduate stage and graduate stage, the rest of schools vary in their schooling conditions. For example, Nanjing University of Aeronautics and Astronautics, Beihang University as well as Civil Aviation University of China, limited by their own practical conditions, rely on foreign flight training institutions to complete their training sessions. And instead, schools like Beijing PanAm International Aviation Academy Co.,Ltd (shortly PanAm Aviation Academy) and Shenyang Feilong Flight Training School, with self-owned flight training facilities and training guarantee system, can only engage in flying skill training since they are not eligible for academic educations.

Except for a few of airline pilots who originally serve in air force and own considerable flight experiences, student pilots graduated from aviation schools generally own 250 up to 280 flight hours when they start to work for airlines, which is a huge gap in comparison with western countries with well-developed GA industry, particularly the United States. In America, people qualified to be employed by airlines for the job of pilots basically have above 2000h flight experiences and most of these experiences are gained either in flight clubs or flight schools, because in the country, there's slim chance for flight students to work in Airlines after graduation, especially when they need to accumulate at least 2000 flight hours for small aircraft and a certain amount of flight time for double engine and medium-sized aircraft so as to be eligible to fly large planes in large companies. Therefore, flight students after graduation may apply to stay in school as instructors or go to the flight club for further training. Such training model, with an aim of taping pilot's full potential, can not only help pilots accumulate flight hours, but also improve their knack for spatial orientation, attitude judgment, flight control, deviation correction and psychological endurance, etc. However, in China, due to the great demand for pilots and the fact that the business of pilot training organizations is still in its infancy, consequently there're no surplus time and available opportunity for such retraining.

As the shortage of civil pilots is a well established fact in China's aviation sector and the current model of pilot training in China can hardly satisfy the market's requirements, the imbalance between pilot's supply and demand will, as a result, become more and more prominent for some time to come, which will then severely affect the normal and safe operation of air transport, set limits on the growth and expansion of Chinese civil aviation and drag down national economy. All in all, how to eliminate the bottleneck of pilot shortage stands out as an urgent issue to be addressed in the development of Chinese civil aviation.

## General Vocabulary

general　*adj.* 一般的，普通的；综合的；大体的
spur　*v.* 激励；促进
industry　*n.* 行业，工业；实业公司
private　*adj.* 私人的，私有的；私立的；私营的
interpretation　*n.* 解释；翻译；
administration　*n.* 管理；行政；实施；行政机构
regulation　*n.* 规则；法规；规章
pilot　*v.* 驾驶；领航
skill　*n.* 技能；技巧；技术
obtain　*v.* 得到
qualify　*v.* 合格的；有资格的；获得资格
organize　*vi.* 组织起来；成立组织
various　*adj.* 各种各样的；多方面的
requirement　*n.* 要求；调整需要量，必需品

standard　*n.* 标准；规格
qualification　*n.* 资格证书；任职资格；职位要求；限定性条件
examination　*n.* 考试；检查；查问
significant　*adj.* 重大的；有效的；有意义的；值得注意的；意味深长的
legalize　*vt.* 使合法化；公认；法律上认为……正当
establish　*v.* 建立；确立
qualification　*n.* 资格；条件
relevant　*adj.* 相关性；有关的；相关的
federal　*adj.* 联邦的；同盟的
subchapter　*n.* 分章
instructor　*n.* 教练；讲师；指导书
curriculum　*n.* 课程
relatively　*adv.* 相当地；相对地，比较地
ordinary　*adj.* 普通的；平凡的；平常的
eligible　*adj.* 合格的，合适的；符合条件的；有资格当选的
undertake　*v.* 同意；担任；许诺
recruitment　*n.* 招收，招聘
candidate　*n.* 候选人，候补者；应试者
understandably　*adv.* 可理解地
comparative　*adj.* 比较的；相当的
cultural　*adj.* 文化的；教养的
academic　*adj.* 学术的；理论的；学院的
literacy　*n.* 识字；读写能力；有文化
obtain　*v.* （尤指经努力）获得；通用
qualification　*n.* 资格证书；任职资格
candidate　*n.* 候选人，候补者；应试者
comparatively　*adv.* 比较地；相当地
academic　*adj.* 学术的；理论的；学院的
education　*n.* 教育；培养；教育学
vocational　*adj.* 职业的，行业的
substantial　*adj.* 大量的；实质的；内容充实的
complete　*v.* 完成；使完整
theoretical　*adj.* 理论的；理论上的；假设的；推理的
correspond　*adj.* 相当的，相应的；一致的；通信的
classify　*v.* 把……分类
irrespective　*adj.* 不考虑的，不顾的
eligible　*adj.* 合格的，合适的；符合条件的；有资格当选的
implementation　*n.* 实现；履行；安装启用
practiced　*adj.* 熟练的；有经验的；老练的
relevant　*adj.* 相关的

annexes  *n.* 附件；附录
roughly  *adv.* 粗糙地；概略地
integrate  *v.* 整合；使……成整体
individual  *adj.* 个人的；个别的；独特的
principle  *n.* 原理，原则
instrument  *n.* 仪器
emerge  *vi.* 浮现；摆脱；暴露
statistics  *n.* 统计；统计学
recreational  *adj.* 娱乐的，消遣的；休养的
barely  *adv.* 仅仅，勉强；几乎不；公开地；贫乏地
contribute  *vi.* 贡献，出力；投稿；捐献
domestic  *adj.* 国内的
lifelong  *adj.* 终身的
regular  *adj.* 定期的；有规律的
transfer  *adj.* 转让的；调动的
employ  *vt.* 雇佣

# Notes

1. According to the interpretation of Provisions on the Administration of Business Licensing for General Aviation under CAAC's regulation No.CCAR-135, training for aircraft pilot license refers to the flight activities carried out with aircrafts for the purpose of mastering piloting skills and obtaining pilot license, which include any flight for normal teaching purpose, flight in the company of trainer and trainee's solo under the guidance of the trainer, but exclude skilled flight.

— the Administration of Business Licensing for General Aviation 《通用航空经营许可管理规定》(CCAR-135TR-R3)是为了加强对通用航空的行业管理，促进通用航空安全、有序、健康地发展而制定的法规，已经2007年1月25日中国民用航空总局局务会议审议通过，自2007年2月14日起施行。

— carry out  实施，贯彻

译文：按照中国民用航空总局CCAR-135部《通用航空经营许可管理规定》中的解释，航空器驾驶员执照培训，"是指使用航空器，以掌握飞行驾驶技术，获得飞行驾驶执照为目的而开展的飞行活动。包括以正常教学为目的的任何飞行，教官带飞和学员在教官的指导下单飞，但不包括熟练飞行。"

2. Flight training model can exert significant influence on the professional growth of pilots and shall be a key issue to be understood and researched in the flight training industry.

— flight training  飞行训练，飞行培训
— influence on  对……的影响，感化
— key issue  关键议题，主要问题，关键问题

译文：飞行培训模式对飞行员成长具有重要影响，也是飞行培训行业需要重点了解与研究的问题。

3. In the United States, the conditions of establishing elementary flight training institutions, as well as the qualification standards of various pilots, and airline's recruitment conditions for pilots are all based on the relevant provisions of the Federal Aviation Regulations (FAR) issued by the Federal Aviation Administration (FAA) of the Department of Transportation.

- Federal Aviation Regulations (FAR) 《联邦航空条例》是指由美国主管民用航空的政府机构—联邦航空管理局颁布的有关民用航空的二级法。涉及民用航空的各个方面，共分15章70部。由于美国民用航空历史长、规模大、技术先进，因而这部法规比较完全，实行的效果比较好，并具有世界性的影响，许多国家直接引用或加以借鉴来制定本国相应的民用航空法规。
- Department of Transportation 运输部，交通部，交通局，美国交通部

译文：在美国，基础飞行培训机构的设立条件，以及各类飞行员资格标准、航空公司对飞行员岗位招聘条件等，都是依据运输部联邦航空管理局（FAA）颁布的《联邦航空条例》的有关法规进行的。

4. All in all, the oversea civil aviation flight schools have low conditions and can easily obtain the qualification certificate for civil aviation flight training, which then bring about a relatively large number of flight training organizations widely located in countries of strong civil aviation industry like the United States.

- all in all 总而言之，总的说来，头等重要的，毕竟
- qualification certificate 资格证，资格证书，资质证书，鉴定合格证
- bring about 带来，导致，引起，发生

译文：国外民航飞行学校设置的条件较低，民航飞行培训资格的取得也不是很难，这就使得美国等民航发达国家飞行培训机构相对较多，并且分布较广。

5. Pursuant to the provisions of American laws, pilot can be eligible to apply for pilot openings in transportation airlines only if he has performed accumulatively 1500-hour flight with small aircrafts as well as flight of certain hours with double-engined aircrafts and middle-sized aircrafts.

- eligible to 有资格
- apply for 申请，请求
- opening 空缺的职位

译文：根据美国法律规定，只有积累够了1500小时的小型飞机飞行与一定的双发及中型飞机飞行时间的人，才有资格到运输航空公司应聘飞行员岗位。

6. With the specified long flying experiences and so many candidate pilots, the rejections are understandably high in the process of airline pilot recruitment. Hence those who finally make it to be an airline transport pilot have comparatively more outstanding piloting skills as well as cultural and academic competence.

- rejections 淘汰率
- academic competence 学术能力，学术竞争力

译文：由于运输航空飞行员选拔的基数大、飞行经历长、淘汰率高，因此，最后真正成为航线运输飞行员的人，其驾驶技术和文化素质也就相对较高。

7. The vast majority of basic pilot training is not for academic education, but for vocational training. As a matter of fact, pilot training has no substantial difference from driving training in which trainees study traffic laws and basic driving skills of motor vehicle at vehicle driving school so as to get the driving license.

- majority of　大多数；大部分
- vocational training　职业培训，职业训练，职业教育
- driving license　驾照

译文：飞行员基础培训绝大多数都不是学历教育，而是职业技能培训。这与到汽车驾驶学校通过学习交通法规和汽车驾驶基本技能，进而取得汽车驾驶证没有什么实质性区别。

8. Their difference lies in that those flight schools organized under CCAR-141 build a more complete and sound internal quality management system and they can provide continuous trainings from level to level.

- lie in　在于，存在于
- CCAR-141　《民用航空器驾驶员学校合格审定规则》

译文：所不同的是，依据141部管理的飞行学校，其内部质量管理体系较为健全完善，并实施连续的培训。

9. With regard to the implementation of flight training, basically it is taught and practiced by aviation theory session and flying skill training session for all countries around the world as all countries are identically bound by the relevant treaties and annexes of International Civil Aviation Organization (ICAO). In the training of flying skill, the training models in different foreign countries are roughly the same, which all follow the stage of PPL, CPL and IR.

- International Civil Aviation Organization (ICAO)　国际民航组织
- PPL　私人飞行执照（英语：Private Pilot Licence/License，简称PPL，美国称之为Private Pilot Certificate），持有人可以驾驶其私人航空器进行非盈利的飞行活动。
- CPL　商用驾驶员执照（英文：CommercialPilotLicense，简称CPL）是一种允许持有人作为飞机的飞行员，并为他的工作获得报酬的资格。
- IR　仪表等级执照（Instrument）

译文：就培训实施而言，因为同样依据国际民航组织有关条约及其附件，所以，世界上所有的国家都大同小异，基本都是分航空理论知识与飞行技能训练两部分来实施教学、训练。在飞行技能训练方面，国外实施培训的模式基本相同，都是按照私照、商照和仪表等级的程序进行。

10. Differently, European countries, Australia, New Zealand and some other countries use sub-course model, meaning that based on the relevancy of theoretical knowledge and contents of aviation, teachings and examinations are done respectively by each individual course, such as flight principle, aircraft structure, propeller engine and jet engine, aviation meteorology, air navigation, instrument and air traffic control.

- flight principle　飞行原理是飞行技术专业一门专业基础课。这门课程的主要特点是既

有抽象的基础理论，又有指导飞行实践的具体原理和方法。通过本课程的学习，使学生获得空气动力的基础理论知识，了解飞机的基本运动规律和基本操纵原理。

— air traffic control  空中交通管制（ATC）

译文：欧洲和澳大利亚、新西兰等采用分课程模式，根据航空理论知识内容的相关性，分为飞行原理、航空器结构、螺旋桨／喷气式发动机、航空气象、空中领航、仪表、飞行管制等课程，分别实施教学和考试。

11. According to ICAO's statistics, globally, operational flights account for approximately 20% of the total flights, whereas the rest two types of GA flight, i.e. business and recreational flights in total take up more than 50% of that in which China barely contributes any.

— business flight  商业飞行

— recreational flight  娱乐飞行

— barely  adv. 几乎不

译文：国际民航组织资料表明，在世界通用航空三类飞行中，航空作业大约占飞行总量的20％，商务和娱乐飞行占飞行总量的50％以上，我国通用航空在商务和娱乐飞行方面几乎是零。

12. Enterprises need to invest a great deal of money in training pilots first and therefore most of them will attach stings of lifelong service to pilots, which will restricts the free flow of pilots among different companies in China.

— a great deal of  许多，大量的，很多

— free flow  自由流动

译文：由于企业对飞行员前期培训的投资很大，在使用上大多是要求终身服役，由此形成国内各公司之间飞行员人才流动的限制。

# Exercises

Section one: Answer the following questions according to the text.

1. In regard to flight experiences of pilots working for airlines, what are main differences in pilots between the US and China?
2. What is the current situation of pilot training in China?
3. Why do airlines treat pilots as private assets and will not easily transfer them out?
4. What is the demand for pilots in the development of air transport in China?
5. What are differences in basic pilot training sessions between China and the US?

Section two: Match the words or phrases in column A with the definition in column B.

| A | B |
| --- | --- |
| 1. curriculum | a. an attribute that must be met or complied with |
| 2. vocational | b. a state at a particular time |

3. asset   c. qualified for or allowed or worthy of being chosen
4. recruitment   d. an integrated course of academic studies
5. academic   e. of or relating to a vocation or occupation
6. competence   f. a useful or valuable quality
7. eligible   g. the quality of being well qualified physically and intellectually
8. ordinary   h. the act of getting recruits;
9. condition   i. not exceptional in any way especially in quality
10. qualification   j. associated with academia or an academy

**Section three: Translate the following into Chinese.**

1. Moreover, pilots should receive a variety of regular trainings on yearly basis, shortly after they work on their posts.

2. Also, considering the expansion of domestic civil aviation industry, the aircraft procurement will see increasing growth and the lack of pilots will become more intense.

3. In addition, pilots retired from air force and flight cadets independently trained by airlines with foreign pilot training institutions are also sources of airline pilots.

4. For example, Nanjing University of Aeronautics and Astronautics, Beihang University as well as Civil Aviation University of China, limited by their own practical conditions, rely on foreign flight training institutions to complete their training sessions.

5. In America, people qualified to be employed by airlines for the job of pilots basically have above 2000h flight experiences and most of these experiences are gained either in flight clubs or flight schools

6. In China, due to the great demand for pilots and the fact that the business of pilot training organizations is still in its infancy, consequently there're no surplus time and available opportunity for such retraining.

7. Such training model, with an aim of taping pilot's full potential, can not only help pilots accumulate flight hours, but also improve their knack for spatial orientation, attitude judgment, flight control, deviation correction and psychological endurance, etc.

8. With the economic growth and the changes of pilot training system in China, private pilot training will emerge as a promising market.

**Section four: Translate the following into English.**

1. 这种病通常无法医治。
2. 我们和几个邻国签署／缔结了限制有害气体排放的条约。
3. 有些疾病是世代遗传的。
4. 这种辛辣食品是该国南部具代表性的食物。

5. 你交的税用作诸如失业和医疗等方面的福利救济。
6. 这些机构竭力地为移民争取权利和福利。
7. 政府最易受人诟病的是经济政策。
8. 市中心的人口密度已下降。
9. 如果试验计划成功，会有越来越多的家庭享受到新的电视服务。
10. 拥有大量金钱并不能保证一定会幸福。

# Key to Exercises
# 练习答案

# Key to Exercises 练习答案

## 1.1

### Section two
1.d  2.f  3.e  4.a  5.c  6.b  7.h  8.g  9.j  10.i

### Section three
1. 美国通用航空发展到今天，已经与运输航空一起组成了美国最安全、最有效的航空运输系统，维系着民航运输业的总体平衡，编织着美国与世界的交通桥梁网络。

2. 随着经济持续快速发展和城乡居民生活水平的不断提高，预计今后10年间我国通用航空年均增长将达到15%以上。

3. 2008年汶川抗震救灾充分显示了通用航空的重要性。在陆路交通中断的情况下，只能依靠直升机第一时间转运伤员、运送物资。

4. 在美国某些大城市的航空港也有通用航空频繁穿梭的足迹。

5. 英国登记为民用的直升机超过1000架，其中大部分在紧急服务（救护、治安、消防、搜救）的前沿领域扮演必要的角色或为必需的经济任务提供支持。

6. 美国通用航空维系着民航运输业的总体平衡，编织着美国与世界的交通桥梁网络。

7. 通用航空的发展紧密地依赖于机场的布局和空域的畅通，其中任何一项如果存在不合理的限制，都将严重损害通用航空和区域经济的发展。

8. 我国的通用航空还有巨大的发展空间和前景，达到美国的发展程度只是时间问题。

### Section four
1. All contributions, no matter how small, will be much appreciated.

2. Label your suitcases to prevent confusion.

3. The country has been in a very poor economic state ever since the decline of its two major industries.

4. The government's economic policies have led us into the worst recession in years.

5. The bomb killed four soldiers and three civilians.

6. Plant growth is most noticeable in spring and early summer.

7. Electronic publishing is an area of activity that is increasing in size and developing quickly.

8. One of the many benefits of foreign travel is learning how to cope with the unexpected.

9. With the benefit of hindsight (helped by the knowledge learned later) it is easy for us to see where we went wrong.

10. We're engaging the services of a professional administrator.

## 1.2

### Section two
1.e  2.j  3.h  4.g  5.b  6.d  7.f  8.c  9.a  10.i

### Section three
1. 我国低空空域管理改革将由1000米推广到3000米，实现监视空域和报告空域无缝衔接。
2. 我国通用航空的广泛商业应用并未正式启动，这是我国改革开放以来唯一没有开放的大产业。
3. 我国大力鼓励旅游业发展，提出积极发展低空飞行旅游。
4. 各国纷纷制订了长远发展战略、相应政策并投入资金；建立了完善的法规和标准体系并培育了相关专业人才，建立通用航空行业组织机构，使得整个国际低空旅游的发展，形成了众多以低空旅游为核心主题的产业门类。
5. 被联合国教科文组织列为世界自然遗产的巴西伊瓜苏瀑布通过直升机观光一览无余；澳大利亚大堡礁的游客乘坐直升机掠过珊瑚海去潜水、泛舟，享受奇妙感觉。
6. 市场对中国低空空域开放一直抱有较大预期，且从未来改革的趋势来看，发展通用航空已成为管理层的共识。
7. "飞不起来""飞不顺畅"是我国通用航空发展的关键瓶颈。
8. 之前我国低空空域一直处于管制状态，我国低空空域管理改革缓慢，导致通用航空"起飞难，落地难"。

### Section four
1. The debate about food safety has engaged the whole nation.
2. Those were the years of private enterprise (= businesses being run privately, rather than by the government), when lots of small businesses were started.
3. The government has refused to rescue the company from bankruptcy.
4. With the new machines we finally have the capability (= power) to do the job properly.
5. In 1906 an earthquake destroyed much of San Francisco.
6. Of course the company will act to protect its financial interests in the country if war begins.
7. Officials will ensure that the election is carried out fairly.
8. She lays the blame for the recession fairly and squarely on the government.
9. Our local shop has very convenient opening hours.
10. Government sources estimate a long-term 50 percent increase in rail fares.

# 1.3

### Section two
1.f  2.g  3.a  4.b  5.h  6.c  7.e  8.i  9.d  10.j

### Section three
1. 民用航空机场是指专供民用航空航空器起飞、降落、滑行、停放以及进行其他活动使用的划定区域，包括附属的建筑物、装置和设施。民用航空机场分为通用航空机场和公共运输机场；不包括临时机场和专用机场。

2. 通用航空机场除了跑道的长度宽度比较小，甚至没有跑道，只有供直升机起降的停机坪。

3. 通用航空机场则开展飞行员培训、空中巡查、空中测绘、气象探测、防林护林、喷洒农药等作业飞行，以及应急救援、商务包机、空中摄影、景点观光、空中表演、私人飞行、短途运输、航空运动等民生功能。

4. A类通用机场，即对公众开放的通用机场，指允许公众进入以获取飞行服务或自开展飞行活动的通用机场。

5. 一个航空项目发展10年后给当地带来的效益，投入产出比可达1∶80，通用航空的就业带动比是1∶12。美国约有2000多个具有通航业务的通用机场，23万架通用航空器，1.9万座通用机场、3500家FBO、1万余家MRO、61个AFSS及开放空域，美国人像开私家车一样翱翔蓝天。

6. 对公众开放的公共性通用航空机场建设应明确其公共基础设施的根本属性，避免盲目贪大求全，从运行实际出发，合理确定其建设规模，按照"一张规划管到底"的要求，逐步发展壮大，等待时机成熟，可升级为运输机场。

7. 根据最新颁布的《通用机场分类管理办法》，准确界定机场的公共性和非公共性定位。

### Section four
1. The number of people who applied for the course was 120 compared with an initial estimate of between 50 and 100.

2. My schedule is flexible – I could arrange to meet with you any day next week.

3. They expanded their retail operations during the 1980s.

4. The city's transport system is one of the most efficient in Europe.

5. We need someone really efficient who can organize the office and make it run smoothly.

6. It's an extremely effective cure for a headache.

7. The foundations will have to be reinforced to prevent the house from sinking further into the ground.

8. There has been a consistent improvement in her attitude.

9. What the witness said in court was not consistent with the statement he made to the police.

10. I was impressed that she found the way here, considering she'd only been he-

re once before.

## 1.4

**Section two**

1.f  2.g  3.a  4.h  5.b  6.c  7.j  8.d  9.e  10.i

**Section three**

1. 2006年通用航空年作业飞行小时数达到98700h。按国际民航组织统计，2006年我国通用航空飞行180000h。

2. 70年代后期至80年代中期，通用航空作业重点是适应国民经济建设中能源勘探和开发的需求，其最有代表性的项目为海上石油服务和陆上石油服务。

3. 随着国家企事业单位购置自用公务机（含直升机）数量的不断增加，对直升机摆渡乘客，直升机医疗救护，以及对小型飞机用于近距离城市之间的旅客穿梭飞行的需求也已凸显。

4. 截至2006年底，全国从事通用航空经营活动的企业已达69家，已批准筹建通用航空的企业30家，全行业从业人员近8000人。比1995年以前通用航空业严重萎缩时的29个作业单位、4000多从业人员的情况有了较大程度的改观。

5. 在通用航空活动的主体形式在原来的基础上，增加了非经营性通用航空活动单位和个人等形式。

6. 虽然通用航空的市场和收入有了一定的发展，但是，在中国民航业总体份额中仍处于相当低的位置。

7. 通用航空开展较好的地区是广东（主要是海上石油服务）、新疆（陆上石油和农业服务）和黑龙江（农林牧业）。

**Section four**

1. The fire caused considerable damage to the church.

2. To conserve electricity, we are cutting down on our heating.

3. There is a strong need to establish effective communication links between staff, parents, pupils, and external bodies.

4. After three months we were well established at our new jobs.

5. For the experiment to be valid, it is essential to record the data accurately.

6. This leaflet will give you the essentials of how to prepare for a hurricane.

7. Criticism just undermines their confidence.

8. Underneath that shy exterior, she's actually a very warm person.

9. Education should be relevant to the child's needs.

10. For further information, please refer to the relevant leaflet.

# 2.1

### Section two
1. e  2. f  3. g  4. a  5. b  6. j  7. i  8. c  9. h  10. d

### Section three
1. 航空小镇以发达的通航产业做支撑，以浓郁的通航文化做基础，建立起适合通航发展的生活方式。

2. 机场目前拥有一条1214米跑道，2家FBO，600余栋带有机库并与机场跑道相连的别墅，一个锦标赛级高尔夫球场，多家高级餐厅和俱乐部。

3. 航空小镇最早起源于美国，充足的机场资源和数量众多的飞行员基础促进了美国航空小镇的出现和成熟发展。

4. 与住宅型航空小镇相比，带有浓厚的工业色彩。而世界各国的知名旅游区分布着大部分的旅游生态小镇。

5. 成熟的通航产业和庞大的通航人口使飞行成为一种日常生活方式，广袤的大型地块和特定的生活习惯促使小镇落成成熟。

6. 随着通用航空的发展以及私人飞行的起步，航空小镇的概念也逐渐被国内认知，很多城市都计划和建设航空小镇。

7. 我国大多数航空小镇是围绕已经建成的通用机场进行规划和建设的。另一部分则是以原有的航空制造产业为基础建设航空产业综合体，并围绕综合体配套相关基础设施。

### Section four
1. You place too much reliance on her ideas and expertise.

2. She has considerable expertise in French history.

3. What is the expiry date of your credit card? (=What is the last date on which it can be used?)

4. Application forms vary greatly in layout and length.

5. We had a four-hour layover in Chicago.

6. There's not much in the way of entertainment in this town – just the cinema and a couple of pubs.

7. This season's entertainments include five new plays and several concerts of Chinese and Indian music.

8. The government has restricted freedom of movement into and out of the country.

9. Growth in car ownership could be restrained by increasing taxes.

10. He has three dependent children.

## 2.2

### Section two
1.e  2.f  3.g  4.a  5.i  6.j  7.b  8.c  9.d  10.h

### Section three
1. 飞机有一个最大的好处之一就是节省时间。飞机出行的独特之处是因为它是基于飞行的交通方式，可以让乘客和货物在两点之间以更直接便捷的方式到达。

2. 众所周知，沃尔玛是全球最大的零售商，2016年年收入高达4821亿美元。如果把沃尔玛当作一个国家，它的GDP全球排28位，拥有一个比南非、阿拉伯联合酋长国和很多其他国家都要大的经济体。

3. 作为一个飞行员，保持身体和心理上的高水准状态是很必要的，有助于进行操作和保持一个良好的判断力。经营一个企业同样需要这些。

4. 自从山姆沃尔顿在1957年买了第一架商务机开始，沃尔玛已经开始培养自己作为全球最大公务机机队之一的自己的公司机队。山姆沃尔顿热爱飞行，他懂得如何利用飞机作为公司的商务战略之一。

5. 公司机队平均每年飞行时间是20000小时，74个飞行员每月平均飞行55个小时。

6. 这些二手飞机在几年后仍然保值。所以沃尔玛购置的这些公务机大多是二手的，不仅是他们企业的飞行文化是不要超支，更是因为公司一直注重产品的低价。

### Section four
1. It's very easy to become dependent on sleeping pills.
2. Whether I get into college or not is dependent on how good my grades are.
3. The council is encouraging the development of the property for both employment and recreation.
4. They've always encouraged me in everything I've wanted to do.
5. The festival is to encompass everything from music, theatre, and ballet to literature, cinema, and the visual arts.
6. It frustrates me that I'm not able to put any of my ideas into practice.
7. Home cooking seems to be on the/in decline.
8. They declined to tell me how they had obtained my address.
9. Illness prevented her attending the launch party for her latest novel.
10. The launch of the space shuttle was delayed for 24 hours because of bad weather.

# 2.3

### Section two
1.h  2.j  3.g  4.a  5.f  6.d  7.e  8.c  9.i  10.b

### Section three
1. 全美国有19100个机场，在FAA注册的约为17800个，公共航空运输仅在其中的651个机场运作，剩余的近18400个机场由通用航空运营或是私人机场。

2. 由政府、团体拥有的机场具有持久和稳定性，特别是由美联邦资金援助兴建的机场更可放心使用。

3. 今天，美国通用航空飞行如此快速、便捷，主要源于美国发达的空中导航网络，这些网络的建立和发展经历了80年的时间。

4. 政府投资安装网络设备，为飞行员提供全程导航。

5. 空中交通管制系统对通用航空飞行至关重要，它监控飞机的相互避让和顺序地在机场终端区域飞进飞出。

6. 通用航空执行航空承运任务，特别是固定的通勤飞行，无论天气好坏，一直都在仪器监控之下。

7. 10年内让小飞机实现以4倍于汽车的速度奔驰于高速公路，到达全国25%的城镇、乡村和偏远地区的目的地；在25年内，到达90%的社区。

8. 1992年提出"雏鹰计划"，目标是到2003年12月航空百年之际，引领100万名青年人进入飞行的世界。

### Section four
1. The airline will launch its new transatlantic service next month.

2. As we approached the hall we could hear the sound of laughter.

3. Advertising companies are always having to think up new ways to promote products.

4. Greenpeace works to promote awareness of the dangers that threaten our planet today.

5. She's just been promoted to senior sales representative.

6. It's a great restaurant but it doesn't look at all promising from the outside.

7. The local residents were angry at the lack of parking spaces.

8. The hotel bar was only open to residents.

9. She is the university's resident expert on Italian literature.

10. There has been a gradual improvement in our sales figures over the last two years.

## 2.4

**Section two**
1.e  2.b  3.h  4.a  5.i  6.g  7.c  8.j  9.f  10.d

**Section three**
1. 公务飞行在通用航空活动中占有很大比例，构成了一个巨大而广泛的公务航空服务体系。
2. 农、林、牧业飞行是美国发展农业的重要手段，每年约有20亿英亩以上的耕地采用通用飞机作业。
3. 2004年，通用航空产业为美国经济直接贡献410亿美元，间接创造价值1020亿美元，为社会提供了近60万个就业机会。
4. 同时，在美国某些大城市的航空港也有通用航空频繁穿梭的足迹。在美国经济建设中，通用航空与运输航空共同为美国的GDP做出了20%的贡献。
5. 南美的巴西作为一个发展中国家，通用航空也较发达，通用飞机达到2万架以上。此外，俄罗斯的通用飞机也有1万架以上。
6. 根据2004年的统计数据，世界上使用中的通用飞机在34万架左右，其中美国约占2/3，约21.1万架。北美洲是全世界通用航空最发达的地区。
7. 20世纪上半叶，世界发生了重大的变化，特别是两次世界大战的爆发，对世界航空技术的发展产生了深远、积极的影响。

**Section four**
1. As you go further south, you will notice a gradual change of climate.
2. We assembled in the meeting room after lunch.
3. The construction work is causing bottlenecks in the city centre.
4. He's had the benefit of an expensive education and yet he continues to work as a waiter.
5. In addition to my salary, I get a pension and medical benefits.
6. The books in the library are classified by/according to subject.
7. I love the classical lines of his dress designs.
8. Part-time work can provide a bridge between staying at home and working fulltime.
9. A German company collaborated with a Swiss firm to develop the product.
10. The British and Italian police collaborated in catching the terrorists.

# 3.1

### Section two
1.d  2.e  3.g  4.a  5.h  6.b  7.j  8.c  9.f  10.i

### Section three
1. 通用航空按照国际惯例，是指除去空军、民航客货运输以外的所有飞行器，飞行内容和飞行活动。

2. 国内保有量200架了，主要用途是飞行执照培训、私人飞行、商用飞行等。

3. H125(AS350B3e)直升机是一款集合坚实耐用、高可靠性、使用成本低等特点的6座单发直升机。

4. H125直升机长度为12.94米，宽2.53米，高3.34米，巡航速度251千米/小时，最大重量2800千克，最大爬升高度5258米，最大航程686千米。

5. R22 BetaⅡ型直升机，以优越的性能，久经考验的安全性，以及实惠的价格和最经济的运营成本，保持在同级别直升机中包括速度、高度、航程的每一项性能记录。

6. 复合动力配件与胜人一筹的飞行操控，可保证多样化环境下的舒适驾乘体验。

7. 具有灵活多变的作业特性，能广泛应用于通航各个领域，可根据任务需要迅速重新构型。

### Section four
1. Anyone who was suspected of collaborating with the occupying forces was arrested.

2. Each union elects several delegates to the annual conference.

3. Authority to make financial decisions has been delegated to a special committee.

4. A group of four teachers were delegated to represent their colleagues at the union conference.

5. Please don't disturb your sister – she's trying to do her homework.

6. Several fans were arrested and charged with disturbing the peace after the game.

7. Some scenes are violent and may disturb younger viewers.

8. She claimed that the way she had been treated at work had caused her extreme emotional and psychological distress.

9. Six people were rescued by helicopter from a fishing boat in distress off the coast.

10. He tried to distract attention from his own illegal activities.

# 3.2

### Section two
1.e  2.j  3.i  4.g  5.b  6.a  7.f  8.c  9.d  10.h

### Section three
1. 目前，庞巴迪宇航集团主要分为庞巴迪公务机部、庞巴迪支线飞机部、飞机服务和新商用飞机项目部、庞巴迪两栖飞机部以及庞巴迪Flexjet和Skyjet公司5个业务部门。

2. 2006财年，庞巴迪宇航集团共交付337架飞机，其中公务机（包括部分拥有项目）和两栖飞机的交付数分别为186架和2架。

3. 湾流公司在不易降低成本的情况下，于1992年开始了超远程喷气公务机"湾流"Ⅴ的研制，该机获得了商业成功，生产了千余架。

4. 1934年，德万·华莱士（Dwane Wallace）在其兄弟德怀特的帮助下接管了该公司，使其逐步发展壮大，在全世界小型通用飞机市场取得了成功。

5. 贝尔直升机德事隆公司是美国著名的直升机兼倾转旋翼机制造商，总部位于得克萨斯州沃斯堡，现为德事隆公司的子公司。

6. 2006年，贝尔直升机德事隆公司销售民用直升机109架，占当年世界总销售量的13.87%。2007财政年度公司销售收入为5.992亿美元，雇员超过8000人。

### Section four
1. I think grey hair on a man can look very distinguished.
2. My original statement has been completely distorted by the media.
3. She's got a very distinctive voice.
4. The union is demanding a seven percent pay rise this year.
5. He has always demanded the highest standards of behaviour from his children.
6. He seems to lack many of the qualities demanded of a successful politician.
7. His new job makes a lot of demands on him.
8. Responsibilities within the department are clearly demarcated.
9. Parking spaces are demarcated by white lines.
10. Financial constraints on the company are preventing them from employing new staff.

# 3.3

### Section two
1.d  2.e  3.j  4.a  5.b  6.c  7.i  8.g  9.h  10.f

## Section three

1. 研究表明，自己拥有飞机的企业，无论是销售量、每股收益、对投资者的长期回报，还是劳动生产率（每个员工的销售额），都比那些没有公务机的企业增长得快。

2. 凡此种种都在提示：公务机运营越来越重要。可以说，航空公司的现状可能是促使人们使用公务机的最好原因。

3. 地球村正在逐渐变成一个国际性的大市场，使用公务机将给这种过渡带来许多方便。

4. 企业业务兴旺时，飞行活动就多；业务萧条时，不仅飞行不多，飞机也会减少，飞行时间和人员随之缩减。

5. 这种兴衰会随着经济状况自有的趋势发生变化，但是公务机运营会持续发展下去，并会随着按需航空运营理念在企业文化中地位的深入而变得更加强大。

6. 实际上，公务机已经成了一种时间机器，它如同《天方夜谭》中的现代化魔毯，能够安全快速地运送管理人员到达下一个商机所在地。

7. 2002年，世界范围内13000多家运营商投放了21000多架燃气涡轮（涡喷和涡桨）式飞机加入公务机运营。

## Section four

1. She tried to appear friendly, but her constraint was obvious.
2. Domestic opinion had turned against the war.
3. These documents are in the public domain (= available to everybody).
4. The commercial future of the company looks very promising.
5. The commentary on the Olympic Games was much better on the other channel.
6. There's good arts coverage in the newspaper, but not much political commentary.
7. Several of the firefighters received commendation for their bravery.
8. He was very proud when his daughter received a commendation for her achievement.
9. We will commence building work in August of next year.
10. For a low-budget film, it has much to commend it.

# 3.4

## Section two

1.d  2.f  3.g  4.i  5.h  6.j  7.b  8.e  9.c  10.a

## Section three

1. 多数集中于经济发达的北美和西欧等地，制造商主要为小型独立公司，通常仅生产很少几种飞机型别及衍生型别。

2. 轻型飞机的发展市场一度衰退，克林顿政府1994年8月12日签署了《通用航空振兴法案》，用立法手段调整通用航空与其他行业的利益关系，迅速扭转了当时美国通用航空下滑的趋势。

3. 下单翼机型根据飞行速度和机翼厚度，起落架分为可收放和不可收放两种，不可收放的起落架基本都配有整流罩。

4. 一般来说，使用的教练机型不一样，收费也不一样；教员的等级不一样，收费也不一样；所学科目不一样，收费也有不同。

5. 目前一共有四种不同类型的飞行员合格证书，最新的一种是娱乐飞行员合格证书，另外三种是私人飞行员合格证书、商业飞行员合格证书以及航线运输飞行员合格证书。

6. 2007年7月22日，赛斯纳飞机公司在奥什科什航展上发布新闻：酝酿1年多的赛斯纳轻型双座运动飞机面世，并开始接受预订单。

7. 该机因其低购买和使用成本，将极富市场前景。前1000架162标准装备飞机的订购价格为10.95万美元/架，首付订金为1万美元。

8. 作为传统产品，单发活塞发动机轻型螺旋桨飞机被用于培养一代又一代飞行员，很多20年前出厂的飞机至今仍在使用。

### Section four

1. A statue has been built to commemorate the 100th anniversary of the poet's birthday.

2. Tears ran down her face as she stood on the winner's podium.

3. You want a plain blouse to go with that skirt – nothing too elaborate.

4. He came out with such an elaborate excuse that I didn't quite believe him.

5. The congresswoman said she was resigning, but refused to elaborate on her reasons for doing so.

6. She's doing the equivalent job in the new company but for more money.

7. Ten thousand people a year die of the disease – that's the equivalent of the population of this town.

8. Yoga is said to restore one's inner equilibrium.

9. The British team have excelled themselves this year to reach the finals.

10. The fall in interest rates is excellent news for borrowers.

# 3.5

### Section two
1. e  2. h  3. f  4. i  5. c  6. g  7. b  8. j  9. d  10. a

### Section three
1. 维修部门的任务就是提供适航的飞机，并在合理成本下尽量有求必应，并负责检查、维护及修理飞机，以确保飞机的安全运行及最大程度的可靠性。

2. 所以，在飞机全寿命周期内，所有维修都不可避免地需要寻求外部援助。目前，关于通用飞机的维修有外包维修和内部维修两种方式。

3. 对某项工作而言，授权的维修机构中的人员资历更深，经验更丰富。

4. 还应详细说明工作的完成情况，以判断所完成工作的类型和数量及工作期间所用部件。最后，对所做工作和所用部件应有质量保证。

5. 此外，在挑选维修商时，要综合考虑他们的资质、名声、工作历史和价格政策。

6. 内部人员应可以完成大多数内部定期维修和不定期维修。

7. 但如果是高水平、非航线的维修任务，就必须选择技术资质与维修任务相符的维修商。

8. 与其他技术人员和维修站人员，特别是那些维修过类似飞机的人员保持联系，通常能为一些技术难题提供解决方案。

## Section four

1. The financial performance of the business is fully expected to improve.

2. She shouldn't be lifting those boxes if she's expecting.

3. I expect punctuality from my students.

4. The president is clearly in a dilemma about/over how to tackle the crisis.

5. Their lawyer was extremely diligent in preparing their case.

6. The discovery was made after years of diligent research.

7. The changes to the national health system will be implemented next year.

8. Shopkeepers are not supposed to sell knives and other sharp implements to children.

9. The owner's name and address is stored on a microchip and implanted in the dog's body.

10. He implanted some very strange attitudes in his children.

# 4.1

## Section two

1.f  2.g  3.h  4.i  5.j  6.d  7.e  8.b  9.c  10.a

## Section three

1. 航空应急医疗救援，就是使用装有专用医疗救护设备的飞机或直升机，为抢救患者生命和紧急施救进行的飞行服务。

2. 据统计，美国目前有100多家专业的空中救护服务运营商，拥有空中医疗救护固定翼飞机（含公务机）超过1000架，每年转运超过70万架次。

3. 此外还要遵守美国联邦航空条例FAR135部中空中救护的相关规定。

4. 目前，美国、英国、德国、法国、意大利、瑞士、日本、挪威、芬兰、澳大利亚等发达国家都已建立较为成熟的航空应急医疗救援体系。

5. 近年来，伴随中国低空空域开放及航空医疗救援体系的不断完善，直升机为主体参与的航空医疗救援逐步进入了大众的视野。

6. 对于航空医疗救援，有些人觉得离自己的生活还很遥远。然而，在危急时刻，这种救援形式却能起到重要作用。

7. 航空医疗救护可以有效节约医疗时间、有效提升医疗处理质量同时有利于后期进一步医疗处理。

8. 最为安全的医疗救护方式。数据证明航空医疗救护的事故率远低于地面救护车救援。

### Section four

1. The whole plot of the film is ridiculously implausible.

2. She accused the party and, by implication, its leader too.

3. The company is cutting back its spending and I wonder what the implications will be for our department.

4. He interpreted her comments as an implicit criticism of the government.

5. The shop said they would replace the television since it was still under guarantee.

6. The label on this bread says it is guaranteed free of/from preservatives (= it contains no preservatives).

7. The fridge is guaranteed for three years.

8. I've always looked to my father for guidance in these matters.

9. The busy traffic entrance was a hazard to pedestrians.

10. Her health has improved dramatically since she started on this new diet.

# 4.2

### Section two
1. e  2. f  3. g  4. h  5. i  6. j  7. d  8. b  9. c  10. a

### Section three

1. 农业航空可以极大地提高劳动生产率，在防御自然灾害和防治有害生物、改善人类生活环境和生态平衡方面发挥重要作用。

2. 农作物持续的高产和稳产依赖于逐步提高管理水平，而作物中后期田间管理水平偏低是目前我国农业生产面临的主要问题。

3. 机械作业极易躺倒或压倒植株，还可能压伤植物根系，在一个较大的区域内很难达到统一防治和迅速控制的目的。

4. 如草地蛾、东亚飞蝗和黏虫等，都曾因为人工防治速度慢而造成极大危害。

5. 航化作业效率高、速度快、病虫害控制明显。此外，航化作业用药量少，最低仅为人工喷洒的十分之一，成本低廉。

6. 同时也能杀死空气中流动的病原菌和害虫，全面抑制病原菌的侵染，控制害虫的蔓延。

7. 除缓和农牧区旱情外，在林区还有降低火险等级和灭火的作用。

8. 我国森林覆盖率较低，因此航空护林的年度作业量相对稳定。

## Section four

1. The governor has denied making improper use of state money.

2. Several companies have been licensed to sell these products.

3. The country's two biggest banks are planning to merge.

4. It was rated the most livable city in the States.

5. The new findings suggest that women ought to monitor their cholesterol levels.

6. The spacecraft will fly around the earth to gain/gather momentum for its trip to Jupiter.

7. The local council has just set up a committee to study recycling.

8. If you're going into teaching, energy is a necessary commodity.

9. English has some features common to many languages.

10. In common with many mothers, she feels torn between her family and her work.

# 4.3

## Section two

1.e  2.f  3.h  4.g  5.j  6.i  7.d  8.b  9.c  10.a

## Section three

1. 我国因发展旅游造成生态破坏屡见不鲜，甚至有些地区发生了不可逆转的生态灾难。

2. 从我国"十二五"开始，通用航空包括农垦喷洒、电力巡护、海上石油、飞行培训以及近年新兴的低空旅游，已经涌现出不少成功案例。

3. 我国大力鼓励旅游业发展，提出积极发展低空飞行旅游。低空旅游作为新兴旅游产品，被纳入国家重点支持的旅游产品。

4. 承租方的经济压力和心理压力很大，绝大部分经营者从经济实力上或是心理承受力上无法将项目坚持一年以上，不少连半年都做不到。

5. 而且部分规定并不透明，造成不少项目在报批阶段就已夭折，因此，有人形容起降点的报批如同走迷宫。

6. 随着经济的发展，加之国外游客和部分国内商务以及中高端游客的需要，此项制约因素已基本弱化消失。

7. 直升机旅游绿色环保，对周围环境影响小；方便快捷节省时间，移位换景受地形条件限制少。

## Section four

1. This brilliant young violinist has appeared on concert platforms all round the world.

2. The US death rate reached a plateau in the 1960s, before declining suddenly.

3. I'd been losing about a pound a week on my diet, but recently I've plateaued

and haven't lost an ounce.

4. A leader with real charisma is needed to revivify the political party.

5. The French Revolution changed France from a monarchy to a republic.

6. No computer can rival a human brain for/in complexity.

7. There's fierce rivalry for the job/to get the job.

8. Among particularly susceptible children, the disease can develop very fast.

9. The dispute is based on two widely differing interpretations of the law.

10. Fortunately the company's bank account is currently in surplus.

## 4.4

### Section two
1.e  2.j  3.g  4.h  5.i  6.d  7.c  8.b  9.f  10.a

### Section three
1. 工业航空在高空进行作业，能够进行人类在地面无法完成的各种生产活动，因而在工业生产建设中发挥着重要作用。

2. 工业航空飞行的首要任务不是将乘客或货物以固定航班的形式从A点运送往B点。

3. 高空作业可以完成人类在地面无法完成的各类生产活动，如航空摄影、航空拍照等。

4. 例如，通用航空在危险性比较大的海上石油开采中的应用，充分发挥了通用航空提供便利快捷的交通、安全的服务以及紧急救援的作用，有效地提高了安全水平。

5. 航空摄影不仅大量用于地图测绘，而且在国民经济建设、军事和科学研究等许多领域中得到广泛应用。

6. 随着飞机和飞行技术，以及摄影机和感光材料等的飞速发展，航空摄影相片的质量有了很大提高，用途日益广泛。

7. 按摄影的实施方式分类，航拍可分为单片摄影、航线摄影和面积摄影。

8. 航空物探作业尤其适用于地形复杂地区的勘查测量工作。

### Section four
1. His time for the 100 metres surpassed the previous world record by one hundredth of a second.

2. To my great surprise, they agreed to all our demands.

3. Not everyone judges success by the same standards—some people think happiness is more important than money.

4. Most people agree that there are standards (of behaviour) that need to be upheld.

5. Her book is still a standard text in archaeology, even though it was written more than 20 years ago.

6. The ferry service has been suspended for the day because of bad weather.

7. The builders worked on wooden platforms, suspended by ropes from the roof of

the building.

8. Of course the ultimate responsibility for the present conflict without doubt lies with the aggressor.

9. Do not miss this unique opportunity to buy all six pans at half the recommended price.

10. We had various problems on our journey, including a flat tyre.

# 4.5

### Section two
1.d  2.e  3.f  4.h  5.j  6.g  7.c  8.i  9.b  10.a

### Section three
1. 此外，飞行员上岗后每年还要进行例行的各项培训。

2. 随着国内民航业的发展壮大，飞机引进越来越多，飞行员需求不足的现象比较明显。

3. 其次是军队航空兵退役飞行员；还有一部分是各航空公司自主与国外飞行员培训机构联合训练的飞行学员。

4. 如南京航空航天大学、北京航空航天大学和中国民航大学，因客观条件的限制，飞行训练阶段需要借助国外的飞行培训机构完成。

5. 因为在美国，飞行学生毕业后要进航空公司工作是很困难的，特别是进大型公司直接飞大飞机需要积累至少2000飞行小时的小型机飞行时间和一定的双发及中型机飞行时间，因此飞行学生毕业后或者申请留校当教员或者去飞行俱乐部进一步训练。

6. 由于我国目前飞行员需求缺口较大，飞行员培训机构尚处于初始状态，没有时间和机会进行这种再训练飞行。

7. 通过这种以挖掘人的最大潜能为目的的飞行训练，不仅积累了飞行小时，还提高了飞行员空间定向、姿态判断、飞行操纵、偏差修正和心理承受能力多方面的锻炼。

8. 随着我国经济的发展和飞行员培养制度的改变，私人飞行员培训成为一个很有发展潜力的市场。

### Section four
1. This disease doesn't generally respond to (= improve as a result of) treatment.

2. We've signed/concluded a treaty with neighbouring states to limit emissions of harmful gases.

3. Some diseases are transmitted from one generation to the next.

4. This kind of hot and spicy food is very typical of the food in the south of the country.

5. Your taxes pay for welfare benefits such as unemployment and sickness pay.

6. These organizations have fought very hard for the rights and welfare of immigrants.

7. It is on economic policy that the government is most vulnerable.

8. The inner cities are no longer densely populated.

9. If the pilot scheme is successful, many more homes will be offered the new television service.

10. The possession of large amounts of money does not ensure happiness.

# Vocabulary
# 词汇表

a set of　一套
abandon　v. 遗弃；离开；放弃；终止；陷入
abnormal　adj. 反常的，不规则的；变态的
abort　vi. 流产；堕胎；夭折；发育不全
absolutely　adv. 绝对地；完全地
absorption　n. 吸收；全神贯注，专心致志
academic　adj. 学术的；理论的；学院的
acceptable　adj. 可接受的；合意的；可忍受的
acceptance　n. 接纳；赞同；容忍
accessible　adj. 易接近的；可进入的；可理解的
accommodate　vt. 容纳；使适应；供应；调解
accomplish　vt. 完成；实现；达到
accordance　n. 按照，依据；一致，和谐
accordingly　adv. 因此，于是；相应地；照着
account for　对……负有责任；对……做出解释；说明……的原因；导致；（比例）占
accumulatively　adv. 累积地
accurate　adj. 精确的
achievement　n. 成就；完成；达到；成绩
acquire　vt. 获得；取得；学到；捕获
acre　n. 土地，地产；英亩
actively　adv. 积极地；活跃地
additionally　adv. 此外；又，加之
address　v. 写（收信人）姓名地址；演说；向……说话，致辞；冠以（某种称呼）；设法解决；就位击（球）；提出
adequate　adj. 充足的；适当的；胜任的
adhere　vi. 坚持；依附；黏着；追随
adjust　vt. 调整，使……适合；校准
administrate　vt. 管理；经营，实施
administration　n. 管理；行政；实施；行政机构
adopt　vt. 采取；接受；收养；正式通过
advanced　adj. 先进的；高级的；晚期的；年老的
advantage　n. 优势；利益；有利条件；vt. 有利于；使处于优势；vi. 获利
advantageous　adj. 有利的；有益的
advent　n. 到来；出现；基督降临；基督降临节
adventure　n. 冒险；冒险精神；投机活动
aerial　adj. 空中的，航空的；空气的；空想的
aerobatics　n. 特技飞行；特技飞行术
aerodynamic　adj. 空气动力学的，航空动力学的
aeroengine　n. 航空发动机；飞机引擎
aeromagnetic　adj. 航空磁测的；空中探测地磁的

aeronautics　*n.* 航空学；飞行术
aerosowing　飞机播种
aerospace　*n.* 航空宇宙；航空航天空间
affection　*n.* 喜爱，感情；影响；感染
affordable　*adj.* 负担得起的
aforementioned　*adj.* 上述的；前面提及的
AFSS　*abbr.* 自动飞行服务站（automated flight service station）
after-sale　售后服务，售后的
agricultural　*adj.* 农业的；农艺的
agricultural reclamation　农垦
agriculture　*n.* 农业；农耕；农业生产；农艺，农学
agrochemical　*adj.* 农用化学品的
agroforestry　*n.* 农用林业；农林复合经营；经济林
aid　*n.* 援助；帮助；助手；帮助者
air freight　航空运费；空运的货物
air show　飞行表演；空中表演
air traffic control　空中交通管制
airborne　*adj.* 空运的；空气传播的；风媒的
aircraft　*n.* 飞机，航空器
Aircraft Owners and Pilot Association　飞机拥有者和飞行员协会
airfield　*n.* 飞机场
airlanding　*v.* （军队或物资等）用飞机降落
airship　*n.* 飞艇
airspace　*n.* 空域；领空；空间
airworthiness　*n.* 适航性；耐飞性
airworthy　*adj.* 适宜航空的；耐飞的；飞机性能良好的
alarm　*n.* 闹钟；警报，警告器；惊慌
alienate from　使疏远，离间；让与
alleviate　*vt.* 减轻，缓和
all-for-one tourism　全域旅游
all-sided　*adj.* 全面的
alternative　*n.* 二中择一；供替代的选择
altitude　*n.* 高地；高度；顶垂线；（等级和地位等的）高级；海拔
ambulance　*n.* 救护车；战时流动医院
amendment　*n.* 修正案；改善；改正
American Association of Airport Executive　美国机场行政人员协会
among　*prep.* 之中；跻身；当中
amphibious　*adj.* 两栖的，水陆两用的；具有双重性的
amusement　*n.* 消遣，娱乐；乐趣
analog　*adj.* 模拟的

ancillary  *adj.* 辅助的；副的；从属的
annexes  *n.* 附件；附录
annually  *adv.* 每年；一年一次
anomaly  *n.* 异常；不规则；反常事物
apart from  远离，除……之外；且不说；缺少
apparatus  *n.* 装置，设备；仪器；器官
apparent  *adj.* 显然的；表面上的
application  *n.* 应用；申请；应用程序；敷用；（对事物、学习等）投入
approach  *n.* 方法，方式；接近；接洽；（某事的）临近；路径；进场（着陆）；相似的事物
approval  *n.* 批准；认可；赞成
approximately  *adv.* 大约，近似地；近于
apron  *n.* 围裙；停机坪；舞台口
Arabian Nights  一千零一夜（书名，又名《天方夜谭》）；不真实的故事
arable  *adj.* 适于耕种的；可开垦的
arguably  *adv.* 可论证地；可争辩地；正如可提出证据加以证明的那样地；可能，大概
arise  *vi.* 出现；上升；起立
arkansas  *n.* （美国）阿肯色州
army worm  黏虫
array  *n.* 数组，阵列；排列，列阵；大批，一系列；衣服
artificial  *adj.* 人造的；仿造的；虚伪的；非原产地的；武断的
as far as  至于，直到，远到；就……而言
assault  *n.* 攻击；袭击
assemble  *vt.* 集合，聚集；装配；收集
assessment  *n.* 评定；估价
asset  *n.* 资产；优点；有用的东西；有利条件；财产；有价值的人或物
assign  *vt.* 分配；指派；赋值
assist  *v.* 参加，出席；协助（做一部分工作）；（通过提供金钱或信息）帮助；在场（当助手）；使便利；（在做某任务中）有助益
assistance  *n.* 援助，帮助；辅助设备
associate  *adj.* 关联的；联合的
association  *n.* 协会，联盟，社团；联合；联想
assume  *vi.* 承担；假定；采取；呈现；装腔作势；多管闲事；*vt.* 假定；僭取；篡夺；夺取；擅用；侵占
assurance  *n.* 保证，担保；（人寿）保险；确信；断言；厚脸皮，无耻
assure  *vt.* 保证；担保；使确信；弄清楚
attach  *vi.* 附加；附属；伴随
attain  *vt.* 达到，实现；获得；到达
attract  *vt.* 吸引；引起
attraction  *n.* 吸引，吸引力；引力；吸引人的事物
attribute  *vt.* 归属；把……归于

## Vocabulary 词汇表

authority　　*n.* 权威；权力；当局
authorized　　*adj.* 经授权的；经认可的
automobile　　*n.* 汽车
autonomous　　*adj.* 自治的；自主的；自发的
auxiliary　　*adj.* 辅助的；副的；附加的；（发动机、设备等）备用的
average　　*adj.* 平均（数）的；普通的；典型的；平庸的；中等的，适中的
Aviation Distributor and Manufacturers Association　　航空经销商和制造商协会
AVIC　　*abbr.* 中国航空工业集团公司（Aviation Industry of China）
avionic　　*adj.* 航空电子学的；航电
awkward　　*adj.* 尴尬的；笨拙的；棘手的；不合适的
backward　　*adj.* 向后的；倒退的；（人）智力迟钝的；落后的；（场地位置）线后的
balloon　　*n.* 气球
bankruptcy　　*n.* 破产
barely　　*adv.* 仅仅，勉强；几乎不；公开地；贫乏地
Barnard　　巴纳德
barometer　　*n.* 气压计；晴雨表；显示变化的事物
Bartoka Gorge　　巴托卡峡
basically　　*adv.* 主要地，基本上
Beaver Lake Aviation　　比弗湖航空
beginner　　*n.* 初学者；新手；创始人
beginning　　*n.* 开始；起点
Belgium　　*n.* 比利时（西欧国家，首都布鲁塞尔 Brussels）
Bell Helicopter　　贝尔直升机
beneficial　　*adj.* 有益的，有利的；可享利益的
benefit　　*n.* 利益，好处；救济金；*vt.* 有益于，对……有益
Bentonville　　*n.* 本顿维尔（美国城市，沃尔玛总部所在地）
biological　　*adj.* 生物的；生物学的
blade　　*n.* 叶片；刀片，刀锋；剑
blindly　　*adv.* 盲目地；轻率地；摸索地
blocked　　*adj.* 堵塞的；被封锁的
blueprint　　*n.* 蓝图；行动方案；（生物细胞的）模型
Boeing　　波音公司
boil down to　　归结起来是其结果是归结为归根结底
Bombardier Aerospace Group　　庞巴迪宇航集团
boost　　*vt.* 促进；增加；支援
bottleneck　　*n.* 瓶颈；障碍物
bottom　　*n.* 底部；末端；臀部；尽头
bought out　　买下……的全部产权
bound to　　必然；一定要
brain child　　*n.* 脑力劳动的产物

brand　　*n.* 品牌，商标；类型；烙印；（独特的）个性；烙铁；污名；燃烧的木头；（诗、文中的）剑

brand-new　　崭新的；全新的

bridge　　*vt.* 架桥；渡过

bruise　　*vt.* 使受瘀伤；使受挫伤

budget　　*n.* 预算，预算费

burgeon　　*v.* 成长；迅速发展；*vi.* 萌芽，发芽；迅速增长

by and large　　大体上，总的来说

CAAC　　*abbr.* 中国民用航空总局（General Administration of Civil Aviation of China）

cabin　　*n.* 小屋；客舱；船舱

call upon　　号召；要求；拜访

candidate　　*n.* 候选人，候补者；应试者

cantilever　　*n.* 悬臂

capability　　*n.* 才能，能力；性能，容量

capable　　*adj.* 有能力的；有才干的；容许……的；可以做（某事）的；综合性的；有资格的

capacity　　*n.* 能力；容量；资格，地位；生产力

cardiovascular　　*adj.* 心血管的

carry out　　执行，实行；贯彻；实现；完成

catalyst　　*n.* 催化剂；刺激因素

catastrophe　　*n.* 大灾难；大祸；惨败

category　　*n.* 种类，分类；范畴

cater to　　迎合；为……服务

CCAR　　*abbr.* 中国民航规章（China Civil Aviation Regulations）

celebrity　　*n.* 名人；名声

cement　　*n.* 水泥；接合剂；纽带；使人们团结的因素；黏固粉；牙骨质；沉积岩基质

centennial　　*adj.* 一百年的

cerebrovascular　　*adj.* 脑血管的

certain　　*adj.* 某一；必然的；确信；无疑的；有把握的

certificate　　*n.* 证书；文凭，合格证书；电影放映许可证

certification　　*n.* 证明，鉴定；出具课程结业证书，颁发证书

Cessna Aircraft Company　　赛斯纳飞机公司

challenge　　*n.* 挑战；怀疑

challenging　　*adj.* 挑战的；引起挑战性兴趣的

characteristic　　*n.* 特征；特性；特色

characteristically　　*adv.* 典型地；表示特性地

characterize　　*vt.* 描绘……的特性；具有……的特征

characterized　　*adj.* 以……为特点的

charge　　*n.* 费用；电荷；掌管；控告；命令；负载

charter　　*v.* 包租；发给特许执照

chemical agent　　化学剂；化学药剂

chemical　　*n.* 化学制品，化学药品
chronicle　　*n.* 编年史，年代记；记录
citation　　*n.* 引用，引证；传票；褒扬
Civil Aviation Flight University of China (CAFUC)　　中国民航飞行学院
civil aviation　　民用航空
civilian　　*adj.* 民用的，百姓的，平民的
claim　　*v.* 宣称；要求，索取；引起（注意）；获得；夺去（生命）；索赔（钱财）；需要
clarification　　*n.* 澄清，说明；净化
classic　　*adj.* 经典的；古典的，传统的；最优秀的；*n.* 名著；经典著作；大艺术家
classification　　*n.* 分类；类别，等级
classify　　*v.* 把……分类
clear up　　清理；放晴；整理；打扫
clear-cut　　*adj.* 清晰的；轮廓鲜明的
climatic　　*adj.* 气候的；气候上的；由气候引起的；受气候影响的
climbing　　*n.* 爬山；攀登；上升
cluster　　*n.* 群；簇；丛；串
coach　　*n.* 教练；旅客车厢；长途公车；四轮大马车
coastal　　*adj.* 沿海的；海岸的
coat　　*vt.* 覆盖……的表面
cockpit　　*n.* 驾驶员座舱；战场
collaboration　　*n.* 合作；勾结；通敌
colleague　　*n.* 同事，同僚
collectively　　*adv.* 集体地，共同地
Colorado Canyon　　科罗拉多大峡谷
combination　　*n.* 结合；组合；联合；化合
comfort　　*n.* 安慰；舒适；安慰者
comment　　*vt.* 发表评论；发表意见
commentator　　*n.* 评论员，解说员；实况播音员；时事评论者
commercial　　*adj.* 商业的；营利的；靠广告收入的
commercialization　　*n.* 商品化，商业化
commodity　　*n.* 商品，货物；日用品
community　　*n.* 社区；群落；共同体；团体
commuter　　*n.* 通勤者，经常乘公共车辆往返者；月季票乘客
comparative　　*adj.* 比较的；相当的
comparatively　　*adv.* 比较地；相当地
compatible　　*adj.* 兼容的；能共处的；可并立的
compensator　　*n.* 补偿器；自耦变压器；赔偿者；补偿物
competent　　*adj.* 胜任的；有能力的；能干的；足够的
competitive advantage　　竞争优势
competitor　　*n.* 竞争者，对手

complement　vt. 补足，补助
complete　v. 完成；使完整
complex　adj. 复杂的；合成的
complexity　n. 复杂，复杂性；复杂错综的事物
compliance　n. 顺从，服从；符合；屈从；可塑性
complicated　adj. 难懂的，复杂的
comply　vi. 遵守；顺从，遵从；答应
component　n. 组成部分；成分；组件，元件
comprehensive　adj. 综合的；广泛的；有理解力的
comprehensively　adv. 包括地；包括一切地
compressor　n. 压缩机；压缩物；收缩肌；压迫器
comprise　vt. 包含；由……组成
compromise　vt. 妥协；连累
concentrate　vi. 集中；浓缩；全神贯注；聚集
conclude　vt. 推断；决定，作结论；结束
conducive　adj. 有益的；有助于……的
conduct　v. 组织，实施，进行；指挥（音乐）；带领，引导；举止，表现；传导（热或电）
configuration　n. 配置；结构；外形
configure　vt. 安装；使成形
confine　vt. 限制；禁闭；（因疾病、残疾）无法离开（床、家、轮椅）；将（自己或自己的活动）局限于；（妇女）卧床分娩
confusion　n. 混淆，混乱；困惑
connotation　n. 内涵；含蓄；暗示，隐含意义；储蓄的东西（词、语等）
consensus　n. 一致；舆论；合意
consequently　adv. 因此；结果；所以
conservation　n. 保存，保持；保护
considerably　adv. 相当地；非常地
consist of　由……组成；由……构成；包括
consolidate　vt. 巩固，使固定；联合
constituent　adj. 构成的；选举的；有任命（或选举）权的；立宪的；有宪法制定（或修改）权的
constitute　vt. 组成，构成；建立；任命
construction　n. 建设；建筑物；解释；造句
constructive　adj. 建设性的；推定的；构造上的；有助益的
consult　vt. 查阅；商量；向……请教
consultation　n. 咨询；磋商；会诊；讨论会
consume　vt. 消耗，消费；使……着迷；挥霍
contentment　n. 满足；满意
continuous　adj. 连续的，持续的；继续的；连绵不断的
contractual　adj. 契约的，合同的

contribute　　vi. 贡献，出力；投稿；捐献
contribution　　n. 贡献；捐献；投稿
convenient　　adj. 方便的；适当的；近便的；实用的
conventional　　adj. 符合习俗的，传统的；常见的；惯例的
coral sea　　珊瑚海
corporate　　adj. 法人的；共同的，全体的；社团的；公司的；企业的
correlate　　vt. 使有相互关系；互相有关系
corresponding　　adj. 相当的，相应的；一致的；通信的
cost-efficient　　adj. 有成本效益的；合算的
council　　n. 委员会；会议；理事会；地方议会；顾问班子
coverage　　n. 覆盖，覆盖范围；新闻报道；保险范围
create　　vt. 创造，创作；造成
credit　　n. 信用，信誉；贷款；学分；信任；声望
crop　　n. 产量；农作物；庄稼；平头
cropland　　n. 农田；植作物之农地
crucial　　adj. 重要的；决定性的；定局的；决断的
cruciform　　adj. 十字形的；十字架状的
cruise　　n. 乘船游览，游船度假；巡航，巡游
cruise speed　　巡航速度
cruise　　v. 乘船游览；以平稳的速度行驶；巡航，巡游，漫游；开车兜风；轻而易举赢得；猎艳（非正式）
cruising altitude　　（飞机）巡航高度
crystal　　n. 结晶，晶体；水晶；水晶饰品
cultivated　　adj. 用于耕种的；有教养的；栽培的，培植的
cultural　　adj. 文化的；教养的
cumulus　　n. 积云；堆积，堆积物
curb　　n. 抑制；路边；勒马绳
curriculum　　n. 课程
curtain　　n. 幕；窗帘
custom　　n. 习惯，惯例；风俗；海关，关税；经常光顾；[总称]（经常性的）顾客
customize　　vt. 定做，按客户具体要求制造
customized　　adj. 定制的；用户化的
cutback　　n. 减少，削减；情节倒叙
cylinder　　n. 圆筒；汽缸；柱面；圆柱状物
daily　　adj. 日常的；每日的
Dassault Aircraft Company　　达索公司
De Haviland Inc.　　德·哈维兰飞机公司
dealer　　n. 经销商；商人
decade　　n. 十年，十年期；十
decline　　vi. 下降；衰落；谢绝

decoration　n. 装饰，装潢；装饰品；奖章
dedicate　vt. 致力；献身；题献
deem　vt. 认为，视作；相信
delegate　v. 授（权），把……委托给他人；委派……为代表，任命
delight　vt. 使高兴
deliver　vt. 交付；发表；递送；释放；给予（打击）；给……接生
delivery　n. 交付；分娩；递送
demand　n.（坚决的或困难的）要求；（顾客的）需求
demonstrate　vt. 证明；展示；论证
densely　adv. 浓密地；密集地
density　n. 密度
depart　vi. 启程；离开
Department of Defense　国防部
depend on　取决于；依赖；依靠
dependent　adj. 依靠的；从属的；取决于……的
deploy　vt. 配置；展开；使疏开
deposit　n. 存款；押金；订金；保证金；沉淀物
depreciate　vi. 贬值；降价
depressed　adj. 沮丧的；萧条的；压低的
depth　n. 深度；深奥
deregulation　n. 放松管制，解除管制，放宽管制，撤销管制
derived　adj. 导出的；衍生的，派生的
descending　v. 下降；下倾；降临；下（坡，楼梯）；依次递降；（声响）渐低；（感觉，气氛）突然笼罩；交由……继承（descend 的现在分词）
desert　n. 沙漠，荒原；沉闷乏味的境况（或地区），冷清的地方；应得的赏罚（常用复数）
designate　vt. 指定；指派；标出；把……定名为
desirable　adj. 可取的，值得拥有的，令人向往的；引起性欲的，性感的
destruction　n. 破坏，毁灭；摧毁
detail　n. 细节，琐事；具体信息；次要部分；分队，支队；vt. 详述；选派
determine　vt. 决定，确定；判定，判决；限定
devastate　vt. 毁灭；毁坏
developed country　发达国家
development　n. 发展；开发；发育；住宅小区（专指由同一开发商开发的）；显影
devoid　adj. 缺乏的；全无的
diameter　n. 直径
diesel　n. 柴油机；柴油；（俚）健康的身体
differential　adj. 微分的；差别的；特异的
differentiated　adj. 分化型；已分化的；可区分的
digital　adj. 数字的；手指的

dilemma　n. 困境；进退两难；两刀论法
direction　n. 方向；指导；趋势；用法说明
disaster　n. 灾难，灾祸；不幸
disaster relief　救灾
dispersing　n. 分配，分散
distanced　v. (使) 远离, 疏远；领先于 (一匹马)(distance 的过去式及过去分词)
distinctive　adj. 独特的, 有特色的；与众不同的
distress　n. 危难，不幸；贫困；悲痛
distribute　vt. 分配；散布；分开；把……分类
disturb　vt. 打扰；妨碍；使不安；弄乱；使恼怒
dive　v. 潜水, 下潜；跳水；(飞机或鸟) 俯冲；冲，奔，扑；突降，暴跌；迅速将手伸入；假摔
diversion　n. 转移；消遣；分散注意力
division　n. 除法；部门；分配；分割；师 (军队)；赛区
doldrums　n. 低谷, 忧郁；赤道无风带
domestic　adj. 国内的；
dominant　adj. 显性的；占优势的；支配的，统治的
dominate　vt. 控制；支配；占优势；在……中占主要地位
door-to-door　adj. 挨家挨户的；送货上门的
dosage　n. 剂量，用量
down payment　(分期付款中的) 头期款；预付定金
downturn　n. 衰退 (经济方面)；低迷时期
dozen　n. 十二个, 一打
drastic　adj. 激烈的；猛烈的
dressing　n. (拌制色拉用的) 调料；馅, 填料；(保护伤口的) 敷料；穿戴；(精修织物用的) 浆料；肥料
drill　vi. 钻孔；训练
drive　v. 推动；钉；开凿；驱赶；迫使，逼迫；激励；挖掘；猛击 (球)；吹, 刮；猛落；围赶 (猎物)；使 (抽象事物) 发生或发展
driveway　n. 车道
drop off　送下车；减少
droplet　n. 小滴, 微滴
drought　n. 干旱；缺乏
drum up　招徕 (顾客)；竭力争取；纠集；鼓动
dubai　迪拜
due to　由于；应归于
durable　adj. 耐用的, 持久的
dye　v. 染；把……染上颜色；被染色
dynamic　adj. 动态的；动力的；动力学的；有活力的
dynamism　n. 活力；动态；物力论；推动力；精神动力作用

225

EAA  *abbr.* American Experimental Aircraft Association 美国实验飞机协会
early maturity  早熟
earnings per share  每股收益，每股盈余，每股盈利
earthquake  *n.* 地震；大动荡
ease  *v.* 减轻，缓解；小心缓缓地移动；使容易；放松；（使）贬值；（股票价格、利率等）下降，下跌
East Asian migratory locust  东亚飞蝗
ecological  *adj.* 生态的，生态学的
ecological environment  生态环境
economic  *adj.* 经济的，经济上的；经济学的
economic strength  经济实力；经济力量
economical  *adj.* 经济的；节约的；合算的
education  *n.* 教育；培养；教育学
effect  *vt.* 产生；达到目的
effectively  *adv.* 有效地，生效地；有力地；实际上
efficient  *adj.* 有效率的；有能力的；生效的
efficiently  *adv.* 有效地；效率高地（efficient 的副词形式）
elaborate  *vt.* 精心制作；详细阐述；从简单成分合成（复杂有机物）
electric power  电力；电功率
electromagnetic  *adj.* 电磁的
eligible  *adj.* 合格的，合适的；符合条件的；有资格当选的
elsewhere  *adv.* 在别处；到别处
embody  *vt.* 体现，使具体化；具体表达
embrace  *vt.* 拥抱；信奉，皈依；包含
Embraer s.a.  巴西航空工业公司
emerge  *vi.* 浮现；摆脱；暴露
emergence  *n.* 出现，浮现；发生；露头
emergency  *adj.* 紧急的；备用的
emergency rescue  急救
emerging  *adj.* 走向成熟的；新兴的
employ  *vt.* 雇佣
employment  *n.* 使用；职业；雇用
empower  *vt.* 授权，允许；使能够
enable  *v.* 使能够；使成为可能；授予权利或方法；（计算机）启动
enclosure  *n.* 附件；围墙；围场
encompass  *vt.* 包含；包围，环绕；完成
encourage  *vt.* 鼓励，怂恿；激励；支持
endurance  *n.* 忍耐力；耐久性；持续时间
enforcement  *n.* 执行，实施；强制
engage in  从事于（参加）

engagement　*n.* 婚约；约会；交战；诺言；进场（游戏术语）
engine　*n.* 引擎，发动机；机车，火车头；工具
enhance　*vt.* 提高；加强；增加
enormous　*adj.* 庞大的，巨大的；凶暴的，极恶的
enter　*vt.* 进入；开始；参加
enterprise　*n.* 企业；事业；进取心；事业心
entertainment　*n.* 娱乐；消遣；款待
entitle　*v.* 给（某人）权利（或资格）；给……题名；称呼
entitlement　*n.* 权利；津贴
entrance　*n.* 入口；进入
entrepreneur　*n.* 企业家；承包人；主办者
entry-level　*adj.* 入门的；初级的；最低阶层的；适合于初学者的
environment　*n.* 环境，外界
environmental　*adj.* 环境的，周围的；有关环境的
environmentally friendly　保护生态环境的；对生态环境无害的
equipment　*n.* 设备，装备；器材
equivalent　*n.* 对等的人（或事物）；当量
era　*n.* 时代；年代；纪元
ERCOUPE 4150C　ERCOUPE 4150C型飞机（ERCOUPE是最安全的单引擎飞机之一）
erode　*vt.* 腐蚀，侵蚀
establish　*v.* 建立，创立；确立；获得接受；查实，证实
estimate　*v.* 估计，估量；判断，评价
Eurocopter and Agust Westland　欧洲直升机公司和阿古斯特·韦斯特兰公司
evacuation　*n.* 疏散；撤离；排泄
evenly　*adv.* 均匀地；平衡地；平坦地；平等地
eventually　*adv.* 最后，终于
examination　*n.* 考试；检查；查问
excel　*vt.* 超过；擅长
except for　除了……以外；要不是由于
exchange　*n.* 交换；交流；交易所；兑换
exchange　*vt.* 交换；交易；兑换
exclude　*vt.* 排除；排斥；拒绝接纳；逐出
exclusively　*adv.* 仅仅；专有地；唯一地；*adv.* 唯一地，专有地，排外地；作为唯一的（消息）来源
execute　*vt.* 实行；执行；处死
executive　*n.* 主管；行政领导；（政府的）行政部门；执行委员会
exert　*vt.* 运用，发挥；施以影响
exhibition　*n.* 展览，显示；展览会；展览品
exotic　*adj.* 异国的；外来的；异国情调的
expand　*vt.* 扩张；使膨胀；详述

expanse   n. 宽阔；广阔的区域；苍天；膨胀扩张
expansion   n. 膨胀；阐述；扩张物
expatriate   n. 移居国外者，侨民；被流放者
exploration   n. 探测；探究；踏勘
expressly   adv. 清楚地，明显地；特别地，专门地
ex-service   adj. 退役的
extend   vt. 延伸；扩大；推广；伸出；给予；使竭尽全力；对……估价
extension   n. 延长；延期；扩大；伸展；电话分机
external   adj. 外部的；表面的；外用的；外国的；外面的
extinguish   vt. 熄灭；压制；偿清
extraction   n. 萃取；家世；提取
extremely   adv. 非常，极其；极端地
FAA   abbr. 联邦航空管理局（Federal Aviation Administration）
face-to-face   adj. 面对面的；当面的
facilitate   vt. 促进；帮助；使容易
facilitation   n. 简易化；助长；容易
facility   n. 设施；设备；容易；灵巧
fade   vi. 褪色；凋谢；逐渐消失
fairing   n. 整流罩；集市上卖的礼物；酬谢礼品
fairly   adv. 相当地；公平地；简直
fall webworms   美国白蛾
falter   vi. 衰弱；衰退；（嗓音）颤抖；结巴地说；蹒跚；犹豫
fancy   adj. 复杂的；昂贵的；精致的，花哨的；想象的；（食物）优质的；（花）杂色的；（动物）供观赏的
far-reaching   adj. 深远的；广泛的；伸至远处的
favorable   adj. 有利的；良好的；赞成的，赞许的；讨人喜欢的
FBO   以固定机场为基地的飞行活动（fixed-base operation）
feature   n. 特色，特征；容貌；特写或专题节目；vt. 特写；以……为特色；由……主演
federal   adj. 联邦的；同盟的；联邦政府的；联邦制的
feedback   n. 反馈；成果，资料；回复
ferry   n. 渡船；摆渡；渡口
fertilizer   n. 肥料；受精媒介物；促进发展者
festival   n. 节日；庆祝，纪念活动；欢乐
finalize   vt. 完成；使结束
fire alarm   火警；火警警报器
firefighting   n. 消防；防火，打火
fire-protection   消防；消防处
first-aid   adj. 急救用的
fiscal   adj. 会计的，财政的；国库的
fiscal year   会计年度；〔美〕财政年度

fishery  n. 渔业；渔场；水产业

fitful  adj. 一阵阵的；断断续续的；不规则的；间歇的

fivefold  adv. 五倍地；五重地

five-year plan  五年计划

fixed  adj. 确定的；固定的；处境……的；准备好的

fixed-wing  固定翼

fleet  n. 船队，舰队；捕鱼船队；（一国的）海军；（同一机构的）车队，机群；（沼泽地）小河，水道，沟

flexible  adj. 灵活的；柔韧的；易弯曲的

flight club  飞行俱乐部

Florida  n. 佛罗里达（美国东南部的州）

flourish  v. 繁荣，茂盛；茁壮成长，处于旺盛时期；挥舞；炫耀

fluctuate  vi. 波动；涨落；动摇

fluctuation  n. 起伏，波动

flying cadet  飞行学员

flying colors  成功；飘扬的旗帜；胜利

focus on  集中于；集中在

foliage  n. 植物；叶子（总称）

forage  n. 饲料；草料；搜索

Forbes Magazine  《福布斯》杂志

forefront  n. 最前线，最前部；活动的中心；（思考、关注的）重心

foreseeable  adj. 可预知的，可预测的

forestation  n. 造林；森林管理

forestry  n. 林业；森林地；林学

formality  n. 礼节；拘谨；仪式；正式手续

formation  n. 形成；构造；编队

formulate  vt. 规划；用公式表示；明确地表达

fortune  n. 财富；命运；运气

foster  v. 促进；抚育（他人子女一段时间）；收养；把（孩子）交托给养父母

foundation  n. 基础；地基；基金会；根据；创立

founder  n. 创立者，创办者，创建者；（动物）新群体建立者；铸造厂业主（或厂长）；浇铸工，制造金属铸件的人

four-cylinder engine  四缸发动机

fragile  adj. 脆的；易碎的

frame  v. 制订；建造

frequency  n. 频率；频繁

frequent  adj. 频繁的；时常发生的；惯常的

frontier  adj. 边界的；开拓的

frustrate  vt. 挫败；阻挠；使感到灰心

fuel  n. 燃料，供给燃料

fulfill　vt. 履行；实现；满足；使结束（与fulfil同）
full load　满载；满负荷
full-fledged　adj. 全面发展的；经过全面训练的；成熟的；有充分资格的；羽毛生齐的
full-metal　全金属
function　vi. 运行；活动；行使职责；n. 功能
fund　n. 基金；资金；存款
fundamental　adj. 基本的，根本的
fundamentally　adv. 根本地，从根本上；基础地
fungicide　n. 杀真菌剂
furnish　vt. 提供；供应；装备
furthermore　adv. 此外；而且
fuselage　n. 机身（飞机）
galaxy　n. 银河；星系；银河系；一群显赫的人
GDP　abbr. 国内生产总值（gross domestic product）
gear　n. 齿轮；装置，工具；传动装置
general　adj. 一般的，普通的；综合的；大体的
General Aviation (GA)　通用航空
generalize　vt. 概括；推广；使……一般化
generally　adv. 通常；普遍地，一般地
geographical　adj. 地理的；地理学的
giraffe　n. 长颈鹿
glance　n. 一瞥；一滑；闪光；（板球）斜击；辉金属
globally　adv. 全球地；全局地；世界上
GPS　abbr. 全球定位系统（Global Position System）
gradual　adj. 逐渐的；平缓的
grand　adj. 宏伟的；豪华的；极重要的
grasp　n. 抓，握；理解，领会；力所能及，把握；权力，控制
grass moth　草地蛾
grassland　n. 草原；牧草地
gravel　n. 碎石；砂砾
Great Barrier Reef　大堡礁（澳大利亚）
gross weight　毛重，总重量
guarantee　n. 保证；担保；保证人；保证书；抵押品
guidance　n. 指导，引导；领导
guideline　n. 指导方针
Gulfstream Aerospace Corporation　湾流公司
handful　n. 少数；一把；棘手事
hang　v. 悬挂
hangar　n. 飞机库；飞机棚
hardware　n. 计算机硬件；五金器具

harmonious　*adj.* 和谐的，和睦的；协调的；悦耳的
harmonize　*vt.* 使和谐；使一致；以和声唱
harsh　*adj.* 严厉的；严酷的；刺耳的；粗糙的；刺目的；丑陋的
havoc　*n.* 大破坏；浩劫；蹂躏
Hawker Beechcraft Corporation (formerly Raytheon Company)　豪客比奇飞机公司（原雷神公司）
hazard　*n.* 危险，风险；冒险的事；机会；双骰子游戏；（高尔夫球的）球场障碍；（庭院网球用语）可得分区域；（台球）落入袋中
headquarter　*n.* 总部；指挥部；司令部
health care　医疗卫生
height　*n.* 高度；高处
Helicopter Association International　国际直升机协会
helicopter　*n.* 直升机
hence　*adv.* 因此；今后
herbicide　*n.* 除草剂
high wing　上单翼
high-end　*adj.* 高端的；高档的
highland　*adj.* 高原的；高地的
highlight　*v.* 突出；强调；使显著；加亮；着亮彩于（头发）
highway　*n.* 公路，大路；捷径
hippo　*n.* 河马
historical　*adj.* 历史的；史学的；基于史实的
hoist　*v.* 升高；举起
honor　*vt.* 尊敬（与honour同）；给……以荣誉
honorably　*adv.* 体面地；值得尊敬地
horizontally　*adv.* 水平地；地平地
hospitalization　*n.* 住院治疗；医院收容；住院保险（与hospitalization insurance同）
hot-air balloon　热气球
hover　*vi.* 盘旋，翱翔；徘徊
hub　*n.* 中心；毂；木片
human resource　人力资源
hurdle　*vt.* 克服
husbandry　*n.* 饲养；务农，耕种；家政
hydrogeological　*n.* 水文地质；水文地质学
Iguazu Falls　伊瓜苏大瀑布（巴西旅游景点）
immediately　*adv.* 立即，立刻；直接地
immune　*adj.* 免疫的；免于……的，免除的
impact　*n.* 影响；效果；碰撞；冲击力
impart　*vt.* 给予（尤指抽象事物），传授；告知，透露
imperfectly　*adv.* 有缺点地；不完美地；未完成地

impetus　　n. 动力；促进；冲力；动量
implement　　vt. 实施，执行；实现，使生效
implementation　　n. 实现；履行；安装启用
implementer　　n. 实施者；制订人；实现器；实作器
impression　　n. 印象；效果，影响；压痕，印记；感想；曝光
improve　　vt. 改善，增进；提高……的价值
in accordance with　　依照；与……一致
in bulk　　整批，散装；大批，大量
in case of　　万一；如果发生；假设
in conclusion　　总之；最后
in light of　　根据；鉴于；从……观点
in stark contrast　　形成了鲜明的对比
in the first place　　首先，起初，第一
incident　　n. 事件，事变；插曲
include　　vt. 包含，包括
incorporate　　vt. 包含，吸收；体现；把……合并
increasingly　　adv. 越来越多地；渐增地
incur　　v. 招致，遭受；引致，带来……
independent　　adj. 独立的；单独的；无党派的；不受约束的
indicate　　vt. 表明；指出；预示；象征
indice　　n. 指数；标记体
indispensable　　adj. 不可缺少的；绝对必要的；责无旁贷的
individual　　adj. 个人的；个别的；独特的
individualized　　adj. 个人的；有个性的；具有个人特色的
industrial　　adj. 工业的；产业的；工业造型
industry　　n. 行业，工业；实业公司
inefficient　　adj. 无效率的，效率低的；无能的
inevitable　　adj. 必然的，不可避免的
infection　　n. 感染；传染；影响；传染病
infrastructure　　n. 基础设施；公共建设；下部构造
inhibitor　　n. 抑制剂，抗化剂；抑制者
initially　　adv. 最初，首先；开头
injure　　vt. 伤害，损害
in-line　　adj. 一列的，直排的；同轴的；（计算机程序）嵌入的，内嵌的；构成完整连续作业一部分的；（生产过程）顺序连接的
innovation　　n. 创新，革新；新方法
in-person　　亲自；外貌上
input-output ratio　　投入产出比
inroad　　n. 损害；得手；侵犯；消耗
insane　　a. 疯狂的；精神错乱的；荒唐的

# Vocabulary 词汇表

insect　*n.* 昆虫；卑鄙的人
insecticide　*n.* 杀虫剂
inspection　*n.* 视察，检查
install　*vt.* 安装；使就任；设置
install　*v.* 安装
institute　*v.* 实行，建立；授予……职位；提出（诉讼）
institution　*n.* 制度；建立；（社会或宗教等）公共机构；习俗
instructor　*n.* 教练；讲师；指导书
instrument　*n.* 仪器
instruments　*n.* 仪器（instrument 的复数）；工具；乐器
insufficient　*adj.* 不足的；不能胜任的，缺乏能力的
insurance　*n.* 保险；保险费；保险契约；赔偿金
insurer　*n.* 保险公司；承保人
intangible　*adj.* 无形的，触摸不到的；难以理解的
integral　*adj.* 积分的；完整的，整体的；构成整体所必须的
integrate　*v.* 整合；使……成整体
intellectual　*adj.* 智力的；聪明的；理智的
intensive　*adj.* 加强的；集中的；透彻的；加强语气的
intensively　*adv.* 强烈地；集中地
intercontinental　*adj.* 洲际的；大陆间的
interior　*adj.* 内部的，里面的；内位的；内陆的，腹地的；内务的，内政的；心灵的，精神的；本质的
interplay　*n.* 相互影响，相互作用
interpretation　*n.* 解释；翻译
interstate　*adj.* 州际的；州与州之间的
introduce　*vt.* 介绍；引进；提出；采用
investigate　*v.* 调查；研究
irregularity　*n.* 不规则；无规律；不整齐
irreplaceable　*adj.*（因贵重或独特）不能替代的，独一无二的；失掉（或损伤）后无法补偿的
irrespective　*adj.* 不考虑的，不顾的
irreversible　*adj.* 不可逆的；不能取消的；不能翻转的
isolate　*v.*（使）隔离，孤立；将……剔出；（某物质、细胞等）分离；区别看待（观点、问题等）
issue　*vt.* 发行，发布；发给；放出，排出
itinerary　*n.* 旅程，路线；旅行日程
jet aircraft　喷气式飞机
join　*vt.* 参加；结合；连接
jointly　*adv.* 共同地；连带地
journey　*n.* 旅行；行程

judgement(美 judgment) n.判断力；裁判
July n.七月
Kansas n.堪萨斯州（美国州名）
kill vt.杀死；扼杀；使终止；抵消
kind n.种类；adj.温和的；无害的
kingdom n.王国；（某人）占统治地位的地方；管辖范围；（某种特质为重的）领域
kilometer n.千米，公里
kilogram n.千克，公斤
know adj.知道的；闻名的；已知的
kw abbr.千瓦特（kilowatt、kilowatts）
labor productivity 劳动生产率
lag vi.滞后；缓缓而行；蹒跚
lagged 延迟
land expatriation 土地征收，土地征用
land on 降落于
land transportation 陆路运输
land vt.使……登陆；使……陷于；将……卸下
landing gear 起落架；起落装置，着陆装置
landscape n.风景；风景画；景色；山水画；乡村风景画；地形
latter adj.后者的；近来的；后面的；较后的
launch v.发射（导弹、火箭等）；发起，发动；使……下水；开始；起飞
law enforcement 法律的实施；执法机关
lay out 展示；安排；花钱；为……划样；提议
layout n.布局；设计；安排；陈列
leading adj.领导的；主要的
Learjet Corporation 里尔公司
lease n.租约；租期；租赁物；租赁权
leaseback n.售后回租
leased v.出租；租用；租借（lease 的过去式和过去分词）
leasing n.租赁，出租
legal person 法人，法定代表人，法人代表
legalize vt.使合法化；公认；法律上认为……正当
legislatively adv.立法地
leisure n.闲暇；空闲；安逸
length n.长度，长；时间的长短；（语）音长
lessee n.承租人
lessor n.出租人
level vt.使同等；对准；弄平
leverage n.手段，影响力；杠杆作用；杠杆效率
liability n.责任；债务；倾向；可能性；不利因素

liaison  *n.* 联络；（语言）连音
license  *n.* 执照，许可证；特许
lifelong  *adj.* 终身的
lifestyle  *n.* 生活方式
light-duty  *adj.* 轻型的
lightweight  *adj.* 轻量的；给人印象不深的；无足轻重的；比通常重量轻的；浅薄的
liken to  与……相比，把……比做
likewise  *adv.* 同样地；也
limitation  *n.* 局限性；（限制）因素；边界
literacy  *n.* 识字；读写能力；有文化
livable  *adj.* 适于居住的；生活过得有价值的；宜居的；值得一过的；可勉强在一起生活的；足够维持生活的；能对付，可处理（与liveable同）
livelihood  *n.* 生计，生活；营生
living standard  生活水平，生活标准
load capacity  负载能力，载重能力
load factor  负载系数；（客机的）座位利用率
location  *n.* 位置；地点；外景拍摄场地
locust  *n.* 蝗虫，蚱蜢
logbook  *n.* 飞行日志；航海日志
logically  *adv.* 逻辑上；合乎逻辑
long-range  *adj.* （飞机、火箭等）远程的；长期的；远大的
low wing  下单翼
lush  *adj.* 丰富的，豪华的；苍翠繁茂的
luxury  *n.* 奢侈，奢华；奢侈品；享受
magnifier  *n.* 放大镜；放大器
maiden  *adj.* （尤指年纪较大女性）未婚的；处女的；（航行、飞行）首次的；（赛马）从未跑赢过的；（植物）生长期第一年内的
maintain  *vt.* 维持；继续；维修；主张；供养
maintain  *vt.* 维持；维修
maintenance  *n.* 维护，维修；保持；生活费用
majority  *n.* 多数；成年
make up  组成；补足；化妆；编造
management  *n.* 管理人员；管理层
managerial  *adj.* 管理的；经理的
mandatory  *adj.* 强制的；托管的；命令的
maneuverability  *n.* 可操作性；机动性
maneuverable  *adj.* 有机动性的；容易操作的；可调动的
manpower  *n.* 人力；人力资源；劳动力
manufacture  *vt.* 制造；加工；捏造
maritime  *adj.* 海事的；海运；生在沿海

maritime monitor　海洋监测
marketization　n.市场化；自由市场经济化；向自由市场经济转化
massive　adj.大量的；巨大的，厚重的；魁伟的
material　n.材料；材质
mature　adj.成熟的；充分考虑的；到期的；成年人的
maximum　adj.最高的；最多的；最大极限的
meanwhile　adv.同时，其间
measure　n.测量；措施；程度；尺寸；vi.测量；估量
medical　adj.医学的；药的；内科的
membership　n.资格；成员资格；会员身份
mentally　adv.精神上；心理上
merely　adv.仅仅，只不过；只是
merge　vt.合并；使合并；吞没
metallic　adj.金属的，含金属的
meteorological　adj.气象的；气象学的
meteorological observation　气象观测
method　n.方法，方式；研究方法
metropolis　n.大都市；首府；重要中心
microelement　n.微量元素
middle-to-high-end　中高端
mile　n.英里；一英里赛跑；较大的距离
military　adj.军事的；军人的；适于战争的
military aircraft　军用飞机
mineral　adj.矿物的；矿质的
mission　n.使命，任务；代表团；布道
model　n.模型；典型；模范；模特儿；样式
modernized　adj.现代化的
modification　n.修改，修正；改变
momentum　n.势头；动量；动力；冲力
monitor　n.监视器；监听器；监控器；显示屏；班长；v.监视，监听，监督
monoplane　n.单翼机
Montreal　蒙特利尔（加拿大东南部港市）
motion　n.动作；移动；手势；请求；意向；议案
MRO　维修、修理服务（maintenance repair operating）
multiple　adj.多重的；多样的；许多的
multitude　n.大量，多数；群众，人群
municipal　adj.市政的，市的；地方自治的
municipality　n.市政当局；自治市或区
namely　adv.也就是；即是；换句话说
narrow　adj.狭窄的，有限的；勉强的；精密的；度量小的

narrow-body　窄体
national　*adj.* 国家的；国民的；民族的；国立的
National Agricultural Aviation Association　国家农业航空协会
National Association of Flight Trainer　国家飞行教官协会
National Association of State Aviation Officials　国家航空官员协会
National Aviation Training Association　国家航空培训协会
National Business Aircraft Association　国家公务机协会
nationwide　*adj.* 全国范围的；全国性的
navigation　*n.* 航行；航海；导航
navy　*n.* 海军，深蓝色的
NBAA (National Business Aviation Association)　*abbr.* 美国国家公务航空协会
nebulizer　*n.* 喷雾器
necessitate　*vt.* 使成为必需，需要；迫使
negotiation　*n.* 谈判；转让；顺利的通过
network　*n.* 网络；广播网；网状物
nevertheless　*adv.* 然而，不过；虽然如此
Niagara Falls Park　尼亚加拉瀑布公园
niche market　瞄准机会的市场；缝隙市场
noise　*n.* 噪声；响声；杂音
nondestructive　*adj.* 无损的；非破坏性的
non-normal　非正常，非常规的
non-profit　*adj.* 非盈利的；不以盈利为目的的
norway　*n.* 挪威（北欧国家名）
nosedive　*n.* （飞机）俯冲
nose-wheel　前轮；机头前轮；前舱
noticeable　*adj.* 显而易见的，显著的；值得注意的
nozzle　*n.* 喷嘴；管口；鼻
numerous　*adj.* 许多的，很多的
nurture　*vt.* 养育；鼓励；培植
obligation　*n.* 义务；职责；债务
observation　*n.* 观察；监视；观察报告
observe　*v.* 观察；遵守
observer　*n.* 观察者；观测者；遵守者
obstacle　*n.* 障碍；障碍物；阻碍
obtain　*v.* （尤指经努力）获得；通用
occasion　*n.* 时机，机会；场合；理由
occupation　*n.* 职业；占有；消遣；占有期
official　*n.* 官员；公务员；高级职员
offset　*v.* 抵消，弥补；衬托出；使偏离直线方向；用平版印刷术印刷，转印下一页；装支管
on average　平均；普通，通常

on-spot　*adj.* 现场的；当场的
opening-up　*n.* 开放
operation　*n.* 运营；运作；业务操作
operational cost　经营成本；操作费用
opportunity　*n.* 时机，机会
optional　*adj.* 可选择的，随意的
orderly　*adv.* 顺序地；依次地
ordinary　*adj.* 普通的；平凡的；平常的
organization　*n.* 组织；机构；体制；团体
organize　*vi.* 组织起来；成立组织
oriented　*adj.* 以……为方向的；重视……的
originally　*adv.* 最初，起初；本来
outbreak　*n.* （战争的）爆发；（疾病的）发作
outlying　*adj.* 边远的；无关的
outperform　*vt.* 胜过；做得比……好
outrank　*vt.* 地位高于，级别高于；居……之上位
outshine　*vt.* 使相形见绌；胜过；比……更亮；*vi.* 放光
outsource　*vt.* 把……外包
outstanding　*adj.* 杰出的；显著的；未解决的；未偿付的
overheads　*n.* 企业的日常管理费用；杂项开支；一般费用（overhead的复数）
oversee　*vt.* 监督；审查；俯瞰；偷看到，无意中看到
oversight　*n.* 监督，照管；疏忽
overspend　*v.* 超支；过度花费
overwhelm　*vt.* 淹没；压倒；受打击；覆盖；压垮
pace　*n.* 一步；步速；步伐；速度
package　*n.* 包，包裹；套装软件，程序包
painstaking　*adj.* 艰苦的；勤勉的；小心的
palpably　*adv.* 易觉察地；可触知地
panoramic　*adj.* 全景的
paraglider　*n.* 滑翔伞
parameter　*n.* 参数；系数；参量
parking lot　停车场
partial　*adj.* 局部的；偏爱的；不公平的
partially　*adv.* 部分地；偏袒地
particularly　*adv.* 异乎寻常地；特别是；明确地
pass away　逝世；停止
passion　*n.* 酷爱；热情；激情
pastoral　*adj.* 牧师的；牧人的；田园生活的；乡村的
pathogen　*n.* 病原体；病菌
patrol　*vt.* 巡逻；巡查

pave　v.（用石板或砖）铺（路），铺砌，铺设；为某事物铺平道路，创造条件
pay through the nose　花很多钱
payload　n.（导弹、火箭等的）有效载荷，有效负荷；收费载重，酬载；（工厂、企业等）工资负担
payroll　n. 工资单；在册职工人数；工资名单；工资
peak　n. 山峰；最高点；顶点；帽舌
performance　n. 性能；绩效；表演；执行；表现
peripheral　adj. 外围的；次要的；（神经）末梢区域的
permission　n. 允许，许可
permit　n. 许可证，执照
personnel　n. 全体职员；员工；人事部门
pest　n. 害虫；有害之物；讨厌的人
pesticide　n. 杀虫剂
petroleum　n. 石油
pharmaceutica　n. 药物
phase out　使逐步淘汰；逐渐停止
photography　n. 摄影；摄影术
physical　adj. 物理的；身体的；物质的；根据自然规律的，符合自然法则的
physically　adv. 身体上；物理上
pick up　（开车等）接人；捡起
pillar　n. 柱子，柱形物；栋梁；墩
pilot　v. 驾驶；领航
pioneer　n. 先锋；拓荒者
piston　n. 活塞
platform　n. 平台；月台，站台；坛；讲台
plunge　v. 使突然地下落；猛插；骤降；陡峭地向下倾斜；颠簸；跳进；（使）陷入；栽种
poisoning　n. 中毒
pontoon　n. 浮筒；浮码头；浮舟；驳船
popularize　vt. 普及；使通俗化
populate　vt. 居住于；构成人口；移民于；殖民于
population　n. 人口；种群，群体；全体居民；总体
portable　adj. 手提的，便携式的；轻便的
position　v. 定位；放置
positive　adj. 积极的；正的，阳性的；确定的，肯定的；实际的，真实的；绝对的
possess　vt. 控制；使掌握；持有；迷住；拥有，具备
possession　n. 拥有；财产；领地；自制；着迷
potential　n. 潜能，可能性；电势
power　n. 力量，能力；电力，功率，性能；政权，势力；幂
power plant　发电厂；动力装置
powerhouse　n. 精力充沛的人，身强力壮的人；强大的集团（或组织）；强国；权势集

团；权威人士；动力源
practiced  adj. 熟练的；有经验的；老练的
precipitation  n. 沉淀，沉淀物；降水；冰雹；坠落；鲁莽
precondition  n. 前提；先决条件
prediction  n. 预报；预言
preferential  adj. 优先的；选择的；特惠的；先取的
pre-order  vt. 预购；预订
prepare  vt. 准备；使适合；装备；起草
presence  n. 存在；出席；参加；风度；仪态
preset  adj. 预先装置的，预先调整的
press release  新闻稿；通讯稿
pricy  adj. 价格高的；昂贵的
primarily  adv. 首先；主要地，根本上
primary  adj. 主要的；初级的；基本的
primitive  adj. 原始的，远古的；简单的，粗糙的
principal  adj. 主要的；资本的
principle  n. 原理，原则
prior to  在……之前；居先
priority  n. 优先；优先权；优先次序；优先考虑的事
privacy-friendly  保护隐私或隐私服务
private  adj. 私人的，私有的；私立的；私营的
Private pilot license  私人飞行员执照
procurement  n. 采购；获得，取得
production efficiency  生产效率
productivity  n. 生产力；生产率；生产能力
professional  n. 专业人员；行家；专家
profitability  n. 盈利能力；收益性；利益率
profound  adj. 深厚的；意义深远的；渊博的
progress  n. 进步，发展；前进
progression  n. 前进；连续
promote  vt. 促进；提升；推销；发扬
prone to  有……倾向的
propagandize  vt. 宣传；对……进行宣传
propeller  n. 螺旋桨；推进器
property  n. 性质，性能；财产；所有权
proportion  n. 比例，占比；部分；面积；均衡
propose  vt. 建议；打算，计划；求婚
propulsion  n. 推进；推进力
prospect  n. 前途；预期；景色；vi. 勘探，找矿
protect  vt. 保护，防卫；警戒

protection  n. 保护；防卫；护照
protective  adj. 防护的；关切保护的；保护贸易的
prototype  n. 原型；标准，模范
province  n. 省；领域；职权
provision  n. 规定；条款；准备；供应品
psychologically  adv. 心理上地；心理学地
public transport  公共交通；公共交通设施
publicity  n. 宣传，宣扬；公开；广告；注意
publicly  adv. 公然地；以公众名义
publicly  adv. 公开地；公然地
pursuant  adj. 依据的；追赶的；随后的
quadruple  adj. 四倍的；四重的
qualification  n. 资格证书；任职资格；职位要求；限定性条件
qualified  adj. 合格的；有资格的
qualify  v. 合格的；有资格的；获得资格
quicken  vi. 加快；变活跃；进入胎动期
radar  n. 雷达，无线电探测器
radical  adj. 激进的；根本的；彻底的
railway  n.〔英〕铁路；轨道；铁道部门
raise  vt. 提高；筹集；养育；升起；饲养，种植
randomness  n. 随意；无安排；不可测性
range  n. 范围；幅度；排；山脉
real estate  n. 不动产，房地产
real-time  adj. 实时的；接到指示立即执行的
reasonable  adj. 合理的，公道的；通情达理的
reassuringly  adv. 安慰地；鼓励地
rebound  v.（球或其他运动物体）弹回，反弹；（价格、价值等下跌后）回升，反弹；抢（篮板球）；（事件，局势）产生事与愿违的结果
reclaim  vt. 开拓；回收再利用；改造某人，使某人悔改
reclamation  n. 开垦；收回；再利用；矫正
reconnaissance  n. 侦察；勘测（与reconnoissance同）；搜索；事先考查
recreation  n. 娱乐；消遣；休养
recreational  adj. 娱乐的，消遣的；休养的
recruitment  n. 招收，招聘
reflect  vt. 反射；反映
reform and opening-up strategy  改革开放战略
reform  v. 改革，革新；重组；（使）改过自新；（石油炼制）重整
region  n. 地区；范围；部位
regional  adj. 地区的；局部的；整个地区的
regional office  地区办事处，办事处，地区办公室

register  v. 登记；（旅馆）登记住宿；挂号邮寄；表达（意见或情感）；显示（读数）
regular  adj. 定期的；有规律的
regulate  vt. 调节，规定；控制；校准；有系统地管理
regulation  n. 管理；规则；校准
relatively  adv. 相当地；相对地，比较地
relevant  adj. 相关性；有关的；相关的
reliability  n. 可靠性
reliable  adj. 可靠的；可信赖的
relieve  vt. 解除，减轻；使不单调乏味；换……的班；解围；使放心
relocate  vt. 重新安置；迁移
rely  vi. 依靠；信赖
remote  adj. 遥远的；偏僻的；疏远的
rename  vt. 重新命名，改名
render  v. 致使；提供，回报；援助，提交，提出；作出（裁决）；放弃；表达；演奏；翻译；绘制；粉刷；熔化；从（动物身体）提取（蛋白质）；秘密偷渡
renewal  n. 更新，恢复；复兴；补充；革新；续借；重申
renowned  adj. 著名的；有声望的
repellent  n. 驱虫剂
replace  vt. 取代，代替；替换，更换；归还，偿还；把……放回原处
representative  n. 代表；典型；众议员；销售代表
reputation  n. 名声，名誉；声望
repute  vt. 名誉；认为；把……称为
request  vt. 要求，请求
requirement  n. 要求；调整需要量，必需品
rescue  v. 营救，援救；（非正式）防止……丢失；n. 营救，解救，援救；营救行动
research  n. 研究；调查
residence  n. 住宅，住处；居住
resident  n. 居民；（旅店）房客；住院医生，（美国的）高级专科住院实习医生；[美]寄宿制学校学生；（英国派驻半独立国家的）特派代表；驻外特工
residue  n. 残渣；剩余；滤渣
resort  vi. 求助，诉诸；常去；采取某手段或方法
respectively  adv. 分别地；各自地，独自地
respond  vi. 回答；作出反应；承担责任
response  n. 响应；反应；回答
responsibility  n. 责任，职责；义务
responsive  adj. 响应的；应答的；回答的
restraint  n. 抑制，克制；约束
restrict  vt. 限制；约束；限定
restriction  n. 限制；约束；束缚
restructure  v. 改组；重组（困难企业的债务）

resume　vt.（中断后）重新开始，继续；重新回到，恢复（席位，地位或职位）；（停顿或被打断后）继续说；重选，重穿，重新占用
retailer　n. 零售商；零售店
retardant　n. 阻滞剂；抑止剂
retract　vt. 缩回；缩进；取消
retractable　adj. 可缩进的；可收起的；伸缩自如的
revenue　n. 税收收入；财政收入；收益
revitalization　n. 复兴，复苏
revival　n. 复兴；复活；苏醒；恢复精神；再生效
revive　vi. 复兴；复活；苏醒；恢复精神
rhino　n. 犀牛（与rhinoceros同）
rigidly　adv. 严格地；坚硬地；严厉地；牢牢地
rival　n. 竞争对手；可与……匹敌的人；同行者
roadside　adj. 路边的；路旁的
robust　adj. 强健的；健康的；粗野的；粗鲁的
rodent　n. 啮齿目动物（如老鼠等）
rogers Municipal Airport　罗杰斯市机场（阿肯色州）
roll out　推出，铺开，转出
roof　n. 屋顶；最高处，顶部；最高限度
rotate　vi. 旋转；循环
rotational　adj. 转动的；回转的；轮流的
rotor　n. 转子；水平旋翼；旋转体
rotorcraft　n. 旋翼飞机
Rotorua City　罗托鲁瓦市
roughly　adv. 粗糙地；概略地
route　n. 路线，航线；道路，公路；（交通工具的）固定路线；巡访；途径，渠道；（北美）递送路线；用于美国干线公路号码前
rudimentary　adj. 基本的；初步的；退化的；残遗的；未发展的
run-up　n. 助跑；预备阶段；抬高；急剧增长
runway　n. 跑道；河床；滑道
rural　adj. 农村的，乡下的；田园的，有乡村风味的
rushed　v. 赶赴,匆忙,繁忙的,冲进
safeguard　n. 保护；保卫；保护措施；预防措施
safety　n. 安全；保险；安全设备；保险装置；安打
saint laurent　圣洛朗
sales　n. 销售额；销售
satellite　n. 卫星；人造卫星；随从；卫星国家
satisfaction　n. 满意，满足；赔偿；乐事；赎罪
saturate　vt. 浸透，使湿透；使饱和，使充满
scale　n. 规模；比例；鳞；刻度；天平；数值范围

scale up　按比例放大；按比例增加
scattered　*adj.* 分散的；散乱的
scenario　*n.* 方案；情节；剧本；设想
scenery　*n.* 风景；景色；舞台布景
scenic　*adj.* 风景优美的；舞台的；戏剧的
scenic spot　风景区；景点
scientific　*adj.* 科学的，系统的
seaplane　*n.* 水上飞机
seasoned　*adj.* 经验丰富的；老练的；调过味的
secondhand　*adj.* 二手的；旧的；间接获得的；做旧货生意的；*adv.* 间接地；间接听来；以旧货
seeding　*n.* 播种；晶种
seemingly　*adv.* 看来似乎；表面上看来
segmented　*adj.* 分段的
seize　*vt.* 抓住；夺取；理解；逮捕
seldom　*adv.* 很少，不常
self-driven　*adj.* 自行驱动的；自励的
self-regulation　*n.* 自我调节，自律，自我调控
sensing　*n.* 感觉，察觉
separately　*adv.* 分别地；分离地；个别地
service ceiling　飞行高度；实用升限；升高限度
set back　推迟；使……受挫折；把……往回拨
set foot in　踏进；进入；涉足于
severe　*adj.* 严峻的；严厉的；剧烈的；苛刻的
share　*n.* 份，份额；股份；责任，贡献
shareholder　*n.* 股东；股票持有人
shift　*v.* 转移；快速移动；变换；改变观点；推卸（责任）；振作；移位；狼吞虎咽地吃；去除（污迹）；销售，出售；换挡；轮班；含糊其词，拐弯抹角
shortage　*n.* 缺乏，缺少；不足
short-distance　短距离；短程
short-haul　*adj.*（尤指空运）短途运输的
shuttle　*n.* 航天飞机；穿梭；梭子；穿梭班机、公共汽车等
side-by-side　*adj.* 并肩的；并行的
sightsee　*vt.* 观光；游览
signal　*n.* 信号；暗号；导火线
significant　*adj.* 重大的；有效的；有意义的；值得注意的；意味深长的
significantly　*adv.* 显著地；相当数量地
skill　*n.* 技能；技巧；技术
slowdown　*n.* 减速；怠工；降低速度
Small Aircraft Manufacturers Association　小飞机制造商协会

small-scale   *adj.* 小规模的

smoothly   *adv.* 平稳地，平滑地；顺利地，流畅地；平静地；均匀地

solid   *adj.* 固体的；可靠的；立体的；结实的；一致的

solo   *adv.* 单独地

sophisticated   *adj.* 复杂的；精致的；久经世故的；富有经验的

sound   *adj.* 合理的；无损的；有能力的；充足的；彻底的；熟睡的；资金充实的；（非正式）非常棒的；严厉的；健全的

soundly   *adv.* 酣畅地；明智地；牢固地；不错地；严厉地；完全地

South Africa   南非

Soviet Union   苏联 [ 1922—1991 年，首都莫斯科（Moscow），位于欧、亚洲 ]

spacious   *adj.* 宽敞的，广阔的；无边无际的

sparsely   *adv.* 稀疏地；贫乏地

specialize   *vi.* 专门从事；详细说明；特化

specialized   *adj.* 专业的；专门的

specification   *n.* 规格；说明书；详述

specified   *adj.* 规定的；详细说明的

specify   *vt.* 指定；详细说明；列举；把……列入说明书

spectacle   *n.* 景象；场面；奇观；壮观；盛大的演出；（复）眼镜

spectrum   *n.* 光谱；频谱；范围；余象

spotting   *v.* 发现；赏识；（尤指体育、娱乐业）业余观察；（尤指从空中）确定敌人的位置；沾上污渍；生斑；散布；用点装饰；下零星小雨；将（台球）放在置球点上；（比赛或运动中）让……一步；借（钱）给（spot 的现在分词）

spray   *n.* 喷雾，喷雾剂；喷雾器；水沫；*vt.* 喷射

spread out   扩展；铺开

spring up   出现；涌现；萌芽

spur   *v.* 激励；促进

squeeze   *v.* 挤；紧握；勒索；压榨；使挤进；（非正式）向……施加压力；（由于金融或商业）破坏

squirrel   *n.* 松鼠；松鼠毛皮

stability   *n.* 稳定性；坚定，恒心

stand out   突出；站出来；坚持到底；坚决反对

standard   *n.* 标准；规格

standard price   标准价格

startup   *n.* 启动；开办

state-of-the-art   *adj.* 最先进的；已经发展的；达到最高水准的

state-owned   *adj.* 国有的；国营的；州立的

statistically   *adv.* 统计地；统计学上

statistics   *n.* 统计；统计学；统计资料

status quo   现状

stay rod   牵条螺栓，缀条，撑杆，锁定杆

steady　*adj.* 稳定的；不变的；沉着的
steep　*adj.* 陡峭的；不合理的；夸大的；急剧升降的
stir　*v.* 搅拌；微动；(使)活动；激发，打动；(非正式)挑拨；传播
stormy　*adj.* 暴风雨的；猛烈的；暴躁的
strategy　*n.* 策略；战略
stratiform　*adj.* 分层排列的；(矿床)成层的
strengthen　*vt.* 加强；巩固
stretcher　*n.* 担架；延伸器
strictly　*adv.* 严格地；完全地；确实地
stride　*vt.* 跨过；大踏步走过；跨坐在……
string　*v.* 用带系上
structure　*n.* 结构；构造；建筑物
stumble　*vt.* 使……困惑；使……绊倒
stunts　*n.* 绝技(stunt的复数)；惊人表演
subchapter　*n.* 分章
sublease　*n.* 转租
subsequent　*adj.* 随后的
subsidiary　*adj.* 辅助的，次要的；附属的；子公司的；*n.* 子公司；辅助者
subsidiary　*n.* 子公司；辅助者
subsidy　*n.* 补贴；津贴；补助金
substantial　*adj.* 大量的；实质的；内容充实的
substantially　*adv.* 实质上；大体上；充分地
successive　*adj.* 连续的；继承的；依次的；接替的
successively　*adv.* 相继地；接连着地
sudden　*adj.* 突然的，意外的；快速的
suffer from　忍受，遭受；患……病；受……之苦
sufficient　*adj.* 足够的；充分的
sum　*n.* 金额；总数
summarize　*vt.* 总结；概述
superior　*adj.* 上级的；优秀的，出众的；高傲的
supervision　*n.* 监督，管理
supposedly　*adv.* 可能；按照推测；恐怕
surging　*v.* 涌，涌动，汹涌；使强烈感到；(物价等)激增；(电流)浪涌(surge的现在分词)
surplus　*n.* 剩余；顺差；盈余；过剩
surround　*vt.* 围绕；包围
surveillance　*n.* 监督；监视
surveying and mapping　测绘科学与技术，测绘学
survival　*n.* 幸存，残存；幸存者，残存物
susceptible　*adj.* 易受影响的；易感动的；容许……的

suspend　　vt. 延缓，推迟；使暂停；使悬浮

sustain　　vt. 维持；支撑，承担；忍受；供养；证实

sustainable　　adj. 可以忍受的；足可支撑的；养得起的；可持续的

sustaining　　adj. 始终如一的；持续的；一致的

Switzerland　　n. 瑞士（欧洲国家）

symbol　　n. 象征；符号；标志

systematically　　adv. 有系统地；有组织地

tackle　　v. 应付，处理（难题或局面）；与某人交涉；（足球、曲棍球等）抢球；（橄榄球或美式足球）擒抱摔倒；抓获；对付，打（尤指罪犯）

tail-wheel　　尾轮

take it easy　　放轻松；别紧张

take off　　起飞；脱下；离开

take over　　接管；接收

talent　　n. 才能；天才；天资

tanker　　n. 油轮；运油飞机；油槽车；坦克手

tap　　v. 轻敲；装上嘴子；窃听；采用；在树上切口；委任；非法劝说转会

Taravila Mountains　　塔拉威拉山

tax　　n. 税金；重负

taxi　　v.（飞机）滑行；乘出租车；用出租车送

technical　　adj. 工艺的，科技的；技术上的；专门的

technician　　n. 技师，技术员；技巧纯熟的人

technological innovation　　技术革新；工艺革新

technology　　n. 技术；工艺；术语

teleconference　　n. 电话会议；远程会议

temporary　　adj. 暂时的，临时的

tendency　　n. 倾向，趋势；癖好

terrain　　n. 地形，地势；领域；地带

territory　　n. 领土，领域；范围；地域；版图

terror　　n. 惊恐；惊慌失措

Textron　　德事隆公司

Thanks to　　由于，幸亏

The General Aviation Manufacturers Association and regional airline associations　　通用航空制造商协会和区域航空公司协会

the United Kingdom　　英国

theme　　n. 主题；主旋律；题目

theoretical　　adj. 理论的；理论上的；假设的；推理的

theoretically　　adv. 理论地；理论上

thickness　　n. 厚度；层；浓度；含混不清

thorny　　adj. 多刺的；痛苦的；令人苦恼的

thoughtfully　　adv. 沉思地；体贴地，亲切地

thrill-seeker　寻找惊险刺激的人
thriving　adj. 欣欣向荣的，兴旺发达的
Tibet　n. 西藏
tilt　v. 倾斜，翘起；俯仰（摄影机）；以言词或文字抨击
timber　n. 木材；木料
time-critical　adj. 时序要求严格的
timely　adj. 及时的；适时的
timetable　n. 时间表；时刻表；课程表
Toronto　n. 多伦多（加拿大城市）
toss out　扔出
total　vt. 总数达
touch　n. 感觉；触摸
tour group　旅行团；旅游团队
tourism　n. 旅游业；游览
tourist　n. 旅行者，观光客
townlet　n. 小镇；小城镇
traditional　adj. 传统的；惯例的
trainee　n. 练习生，实习生；受训者；新兵；训练中的动物
training　n. 训练；培养；瞄准；整枝
transceiver　n. 收发器，无线电收发两用机
transfer　v. 转让；转接；移交；转移（地方）；（使）换乘；转存，转录；调动（工作）；传染，传播；使（运动员）转队；把（钱）转到另一账户，机构上
transform　vt. 改变，使……变形；转换
transformation　n. 转化；转换；改革；变形
transit　vt. 运送
transmit　vt. 传输；传播；发射；传达；遗传
transportation　n. 运输；运输系统；运输工具；流放
trapezoidal　adj. 梯形的；不规则四边形的
trauma　n. 创伤（由心理创伤造成精神上的异常）；外伤
treatment　n. 治疗，疗法；处理；对待
trip　vi. 绊倒，跌倒；轻快地走（或跑、跳舞）；（非正式）（服用毒品后）产生幻觉；（部分电路）自动断开；作短途旅行
trivial　adj. 不重要的，琐碎的；琐细的
troubleshooter　n. 解决纠纷者；故障检修工；（计算机）故障查找软件
turbojet　n. 涡轮喷气飞机
turboprop　n. 涡轮螺桨发动机；涡轮螺旋桨飞机
turboshaft　n. 涡轮轴；涡轮轴发动机
typhoon　n. 台风
typical　adj. 典型的；特有的；象征性的
UAE　abbr. 阿拉伯联合酋长国（United Arab Emirates）

UAV　*abbr.* 无人机（unmanned aerial vehicle）
ultimate　*adj.* 最终的；极限的；根本的
ultra　*adj.* 极端的，偏激的
ultra light aircraft　超轻型飞机
ultra-micro　超微量
undemanding　*adj.* 要求不高的；容易的；不严格的
underdeveloped　*adj.* 不发达的
undergo　*vt.* 经历，经受；忍受
undermine　*vt.* 破坏，渐渐破坏；挖掘地基
underpin　*vt.* 巩固；支持；从下面支撑；加强……的基础
understandably　*adv.* 可理解地
undertake　*v.* 同意；担任；许诺
undertake　*vt.* 承担，保证；从事；同意；试图
undulate　*v.* 使波动；使成波浪形
uneconomical　*adj.* 不经济的；浪费的；不节俭的
UNESCO　*abbr.* 联合国教科文组织（United Nations Educational, Scientific, and Cultural Organization）
unfavorable　*adj.* 不宜的；令人不快的；不顺利的
uniform　*adj.* 统一的；一致的；相同的；均衡的；始终如一的
uniformly　*adv.* 一致地
unique　*adj.* 独特的，稀罕的；唯一的，独一无二的
United States Ultralight Association　美国超轻型飞机协会
unlikely　*adj.* 不太可能的；没希望的
unparalleled　*adj.* 无比的；无双的；空前未有的
unprecedented　*adj.* 空前的；史无前例的
untapped　*adj.* 未开发的；未使用的；塞子未开的
up and down　上上下下；到处；前前后后；来来往往
upgrade　*v.* 使（计算机、软件等）升级；改善（尤指服务）；给（飞机乘客或宾馆客人）升级；给（某人）升职；提高（某事物的）地位
upper-class　*adj.* 上流社会的；上层阶级的；中学三年级，四年级的
urban　*adj.* 城市的；住在都市的
urbanized　*adj.* 城市化的
urgent　*adj.* 紧急的；急迫的
usher　*vt.* 引导，招待；迎接；开辟
utilization　*n.* 利用，使用
utilize　*vt.* 利用
vacation　*n.* 假期；（房屋）搬出
validate　*vt.* 证实，验证；确认；使生效
various　*adj.* 各种各样的；多方面的
vast　*adj.* 广阔的；巨大的；大量的；巨额的

vegetative　*adj.* 植物的；植物人状态的，无所作为的；促使植物生长的；有生长力的
versatile　*adj.* 多才多艺的；通用的，万能的；多面手的
Victoria falls　维多利亚瀑布
vigorously　*adv.* 精神旺盛地，活泼地
villa　*n.* 别墅；郊区住宅
virtually　*adv.* 事实上，几乎；实质上
visible　*adj.* 明显的；看得见的；现有的；可得到的
vocational　*adj.* 职业的，行业的
vulnerable　*adj.* 易受攻击的，易受……的攻击；易受伤害的；有弱点的
wait-and-see　*adj.* 观望的
walk-in　*adj.* 无预订散客，（宽敞得可以）步入的；（无需、未经）预约的
water-logging　水浸，水渗；浸透
waterway　*n.* 航道；水路；排水沟
weaken　*vt.* 减少；使变弱；使变淡
wealthy　*adj.* 富有的；充分的；丰裕的
weeding　*n.* 除草，除杂草
welfare　*n.* 福利；幸福；福利事业；安宁
wellbeing　*n.* 幸福；福利；安乐
well-developed　*adj.* 发达的；发育良好的
well-known　众所周知的，出名的，知名的
whereas　*conj.* 然而；鉴于
whilst　*conj.* 当……的时候；与……同时；然而；虽然，尽管；直到……为止（与于while同）
whip up　激起；鞭打
wide-body　宽体
widespread　*adj.* 普遍的，广泛的；分布广的
wildfire　*n.* 野火，烈火；希腊火；高度易燃物质，燃烧剂；鬼火
wildlife　*n.* 野生动植物
wing　*n.* 翼；翅膀；飞翔；派别；侧厅，耳房，厢房
wingspan　*n.* 翼展；翼幅
Winter Olympic Games　冬季奥运会
win-win　*adj.* 双赢的；互利互惠的
with regard to　关于；至于
Witness　*vt.* 目击；证明；为……作证
Wooden　*adj.* 木制的；僵硬的，呆板的
World Heritage Site　世界遗产地
World War Ⅰ　第一次世界大战
worthwhile　*adj.* 重要的；令人愉快的；有趣的；值得（花时间、金钱、努力等）的；有价值的，有益的；值得做的
worthy　*adj.* 值得的；有价值的；配得上的，相称的；可尊敬的；应……的

wreak　*vt.* 发泄；报仇；造成（巨大的破坏或伤害）
wright brothers　莱特兄弟（飞机发明者）
WW Ⅱ　*abbr.* 第二次世界大战（World War Ⅱ）
year-on-year　*adj.* 与上年同期数字相比的
yield　*n.* 产量；利润，红利率；*v.* 出产（产品或作物）；产出（效果、收益等）；生息；屈服；放弃；停止争论；给（车辆）让路；（在外力、重压等下）屈曲
Zambezi River　赞比西河（非洲南部河流）
Zambia　*n.* 赞比亚（非洲中南部国）
zone　*n.* 地带；地区；联防

# 相关通用航空知识

## 1. 飞行机构组织

（1）CAAC（Civil Aviation Administration of China）中国民用航空局（简称：中国民航局或民航局，英文缩写CAAC）是中华人民共和国国务院主管民用航空事业的由部委管理的国家局，归交通运输部管理。

（2）FAA（Federal Aviation Administration）联邦航空管理局，是美国一个具有管理民用航空权力的国家机构。它负责机场建设和运营，管理空中交通、人员和飞机认证，以及商业航空器发射或折返，是世界上主要的航空器适航证颁发者。

（3）EAA（Experimental Aircraft Association）实验飞机协会，是美国航空爱好者自发的群众性的飞行大会，每年夏季都隆重举行一次集会，每次都有上万架大小各异的飞机和近百万狂热的美国航空爱好者参加这个飞行盛会，其中绝大多数是私人拥有的或是自制的小型飞机。EAA展现了美国民众强烈的航空意识和航空热情，是美国群众性的航空活动发展到一定规模后的必然产物，同时也显示了美国这个航空大国背后的群众基础，也能从一个侧面说明美国能够成为世界一流航空大国的原因所在。

（4）NBAA（National Business Aviation Association）美国公务航空协会，创立于1947年，总部设于华盛顿特区，它为依靠民用飞机提高效率和生产率并获取成功的公司提供服务，并且是其中翘楚。

（5）AVIC（Aviation Industry Corporation of China）中国航空工业集团有限公司，是中国国家出资设立，由国务院国资委代表国务院履行出资人职责的国有独资公司，是由中央管理的国有特大型企业，也是国家授权投资的机构，于2008年11月6日由原中国航空工业第一集团公司和中国航空工业第二集团公司重组整合成立。中国航空工业集团有限公司设有航空武器装备、军用运输类飞机、直升机、机载系统与汽车零部件、通用航空、航空研究、飞行试验、航空供应链与军贸、资产管理、金融、工程建设、汽车等产业。

（6）Air Methods 创立于1980年，总部位于美国科罗拉多州Englewood，全职雇员4554人，连同其子公司，在美国提供直升机运营，提供全美航空医疗紧急运送服务给各家医院。以独立方式提供航空医疗运输服务于一般民众、医院及签有独家经营合作协议的机构。服务包括飞机的操作和维修、医疗保健、派遣和通讯、医疗计费和收款。截至2015年12月31日，该部门的飞机机队包括321架公司自有的飞机和92架租赁飞机，以及54架客户的飞机，由AMS与他们签订合同经营。

## 2. 飞行执照（Pilot license）

飞机驾驶员需要持有相应执照，规范的相关规则是由美国联邦航空管理局（FAA）制定的，该局为美国交通运输部（DOT）的下属机构。飞行员的认证规定编纂在"联邦法典"（CFR）的第14卷的61部和141部，也被称之为"联邦航空法规"（FARs）。

飞行员根据级别认证的不同可驾驶一种或多种级别的飞机，每种级别的飞机分属于某种经认定的类型。按级别递增，飞行员执照等级如下：

① 学员（Trainee）：在飞行教官监管下学习飞行的个人，在特别受限制的环境下允许单飞。

② 运动类飞行员（Sports pilot）：允许飞行轻型运动飞机的飞行员。

③ 娱乐类飞行员（Recreational pilot）：允许日间飞行180马力以下和4座以下飞机的飞行员。

④ 私照飞行员（Private pilot）：不收取商酬，仅为个人娱乐或个人私务而从事飞行的飞行员。

⑤ 商照飞行员（Commercial pilot）：在某些限制下，可进行取偿飞行或雇佣飞行的飞行员。

⑥ 航线运输飞行员（ATP-Airlines transportation pilot）：被允许在定期航班作为机长的飞行员。

学员驾照在学院的首次身体检查时由航空体检员（AME）签发；对于不需要体格检查的情况，学员驾照由FAA检查员或FAA指定飞行考官。学员驾照仅在进行单飞时需要。学员驾照在签发后（根据年龄的不同）24个月或60个月的最后一天失效。一旦学员积累了有效的培训和飞行经验，注册飞行教官（CFI）可通过签字的方式许可学员单飞某种特定品牌和型号的飞机。学员不可载客，不可进行业务飞行、不可在CFI签发范围外进行操作。

运动飞行员执照经过实验飞机协会（EAA）多年工作后自2004年9月份创制。新规则的初衷是降低航空领域的门槛和使飞行更易承担和进入。

娱乐飞行执照比私照要求较少的培训和提供较少的级别认证。其最初是用来飞行小型单发飞机，现在大部分已经逐步被运动执照所替代。其限制性比私照更大。娱乐执照持有人被局限在起飞地和具有塔台的机场附近50海里的区域内，尽管经过额外的培训和书面认证，该限制可被进一步放松。

私照飞行员：私照被当今大多数在飞的飞行员所持有。该执照允许驾驶任何（满足适当的级别要求下）飞机进行非商业用途的飞行，并且在目视飞行规则下几乎没有限制。可以载客并进行本人商务飞行。然而，私照飞行员不允许以收取报酬为目的提供飞行驾驶服务，尽管乘客可以分担飞行成本，比如燃油和租机开支。在某些限制下，私照飞行员可以进行慈善飞行，也可以参与到类似的活动中，比如"天使飞行""民间巡逻"及其他。

商照飞行员：商照飞行员可以取酬飞行。该执照的培训着重于对飞机系统和制造工艺等方面的较高要求。商照本身并不允许飞行员进行仪表气环条件飞行，对于没有仪表级别的商照飞行员仅允许50海里内的日间飞行。

航线运输飞行员：航线运输飞行员（通常称之为ATP）要经受最高级别飞行能力的测试。航线执照是成为定期航班运营机长的前提条件。

### 3. 通航相关法规

the Administration of Business Licensing for General Aviation《通用航空经营许可管理规定》（CCAR—135TR—R3）是为了加强对通用航空的行业管理，促进通用航空安全、有序、健康地发展而制定的法规，已经2007年1月25日中国民用航空总局局务会议审议通过，自2007年2月14日起施行。

Federal Aviation Regulations (FAR)《联邦航空条例》是指由美国主管民用航空的政府机构——联邦航空局颁布的有关民用航空的二级法。涉及民用航空的各个方面，共分15章70部。由于美国民用航空历史长、规模大、技术先进，因而这部法规比较完全，实行的效果比较好，并具有世界性的影响，许多国家直接引用或加以借鉴来制定本国相应的民用航空法规。

Regulation on the Classification of GA Airport《通用机场分类管理办法》是为贯彻

落实国务院办公厅《关于促进通用航空业发展的指导意见》要求，促进通用机场健康发展制定。由民航局于2017年4月14日印发并实施。《办法》按照通用机场是否对公众开放分为A、B两类。A类为对公众开放的通用机场，允许公众进入以获取飞行服务或自行开展飞行活动；B类则为不对公众开放的通用机场。另外，基于其对公众利益的影响程度，《办法》又将A类通用机场分为三级。其中，含有使用乘客座位数在10座以上的航空器开展商业载客飞行活动的为A1级通用机场，使用座位数在5~9座之间的航空器开展商业载客飞行活动的为A2级通用机场，其余均为A3级通用机场。今后，我国所有通用机场的建设与运行管理都将遵循此《办法》。

### 4. 航空类专业名词

① Aviation petroleum　石油航空，系以直升机为主工具，辅以小型固定翼飞机，为海洋和陆地石油、天然气资源勘探、开发和管理提供空中运输的作业。国际上亦称石油开发后勤支援。石油航空是随石油能源勘探、开发的发展而产生的。它具有迅速、灵活、机动的优点。它已成为石油勘探、开发重要的后勤保证手段。中国在塔里木盆地和沿海海域的石油勘探开发均利用了石油航空。

② Aerial mapping　航空测绘，是一种以大气层内的飞行器为测量载体的对地测绘手段，其测绘对象是地面物体的位置关系，目的是通过航空拍摄获得的数据来绘制大地坐标，其通常采用的方法是航空摄影测量。航空摄影测量是在飞机上利用航摄仪器对地面进行连续拍摄，绘制地形图的过程。

③ Strip photography　航线航空摄影，又称"带区航空摄影"，是沿宽度小而长度大的狭长地带所进行的航空摄影。通常沿一条或两条航线对地面狭长地区或线状地物进行连续摄影。航向相邻相片之间有一定重叠，一般为60%，不少于53%。常用于河流或道路的勘测。

④ Area photography　沿数条航线对较大区域进行连续摄影，称为面积摄影（或区域摄影）。面积摄影要求各航线互相平行，在同一条航线上相邻相片间的航向重叠为60%~53%，相邻航线间的相片也要有一定的重叠，这种重叠称为旁向重叠，一般应为30%~15%。实施面积摄影时，通常要求航线与纬线平行，即按东西方向飞行，但有时也按照设计航线飞行。由于在飞行中难免出现一定的偏差，故需要限制航线长度一般为60~120千米，以保证不偏航，而产生漏摄。

⑤ GPS navigation　全球定位系统导航（Global Position System），利用GPS定位卫星，在全球范围内实时进行定位、导航的系统，GPS是由美国国防部研制建立的一种具有全方位、全天候、全时段、高精度的卫星导航系统，能为全球用户提供低成本、高精度的三维位置、速度和精确定时等导航信息，是卫星通信技术在导航领域的应用典范，它极大地提高了地球社会的信息化水平，有力地推动了数字经济的发展。

⑥ Crash radio　应急广播是指当发生重大自然灾害、突发事件、公共卫生与社会安全等突发公共危机时，造成或者可能造成重大人员伤亡、财产损失、生态环境破坏与严重社会危害，危及公共安全时，应急广播可提供一种迅速快捷的讯息传输通道，在第一时间把灾害消息或灾害可能造成的危害传递到民众手中，让人民群众在第一时间知道发生了什么事情，应该怎么撤离、避险，将生命财产损失降到最低。

⑦ ATC（Air Traffic Control）　利用通信、导航技术和监控手段对飞机飞行活动进行监视和控制，保证飞行安全和有秩序飞行。在飞行航线的空域划分不同的管理空域，包括航路、

飞行情报管理区、进近管理区、塔台管理区、等待空域管理区等，并按管理区不同使用不同的雷达设备。在管理空域内进行间隔划分，飞机间的水平和垂直方向间隔构成空中交通管理的基础。由导航设备、雷达系统、二次雷达、通信设备、地面控制中心组成空中交通管理系统，完成监视、识别、导引覆盖区域内的飞机。

⑧ ILS（Instrument Landing System） 仪表着陆系统，又译为仪器降落系统、盲降系统，是应用最为广泛的飞机精密进近和着陆引导系统。它的作用是由地面发射的两束无线电信号实现航向道和下滑道指引，建立一条由跑道指向空中的虚拟路径，飞机通过机载接收设备，确定自身与该路径的相对位置，使飞机沿正确方向飞向跑道并且平稳下降高度，最终实现安全着陆。仪表着陆系统能在气象条件恶劣和能见度差的条件下向飞行员提供引导信息，保证飞机安全进近和着陆。ILS系统包括3个分系统：提供横向引导的航向信标、提供垂直引导的下滑信标(glide slope)；提供距离引导的指点信标(marker beacon)。每一个分系统由地面发射设备和机载设备组成。

⑨ UAV（Unmanned Aerial Vehicle/Drones） 无人驾驶飞机简称"无人机"，英文缩写为"UAV"，是利用无线电遥控设备和自备的程序控制装置操纵的不载人飞机，或者由车载计算机完全地或间歇地自主地操作。与载人飞机相比，它具有体积小、造价低、使用方便、对作战环境要求低、战场生存能力较强等优点。2020年初，新冠疫情已在多国爆发，许多国家开始学习中国的防疫经验，这其中也包括使用无人机等高科技。不久前，利用中国的大疆无人机测体温、监控、喊话的桥段已经在欧美多国上演。现在，用植保无人机直接消杀作业的场景也出现在了欧洲、美洲、亚洲的多座城市。

⑩ FBO 以固定机场为基地的飞行活动（fixed-base operation），位于机场或者邻近机场的为通用航空飞机、公务机和私人飞机提供停场、检修、加油、清洁、休息等服务的基地或服务商。FBO的服务对象也主要是通用航空飞机，特别是公务机和私人飞机。FBO通航企业性质是综合性的通用航空服务企业或基地。提供服务是停场、检修、加油等服务，业务范围比较广泛。除了飞机的维护、维修外，包括飞机销售、租赁和飞行培训等综合服务。给私人飞机客户提供全方位立体的服务。

⑪ VOR（VHF omnidirectional radio range） 甚高频全向信标是指一种工作于112~118MHz，可在360度范围内给航空器提供它相对于地面台磁方位的近程无线电导航系统，一种用于航空的无线电导航系统。其工作频段为112~118兆赫的甚高频段，故此得名。VOR发射机发送的信号有两个：一个是相位固定的基准信号；另一个信号的相位随着围绕信标台的圆周角度是连续变化的，也就是说各个角度发射的信号的相位都是不同的。向360度（指向磁北极）发射的与基准信号是同相的（相位差为0），而向180度（指向磁南极）发射的信号与基准信号相位差180度。飞行器上的VOR接收机根据所收到的两个信号的相位差就可以计算出自身处于信标台向哪一个角度发射的信号上。VOR通常与测距仪（Distance Measuring Equipment，DME）同址安装，在提供给飞行器方向信息的同时，还能提供飞行器到导航台的距离信息，这样飞行器的位置就可以唯一的被确定下来。

⑫ EMS（emergency medical service） 紧急医疗服务，也称"急救"。当有任何意外或急病发生时，施救者在医护人员到达前，按医学护理的原则，利用现场适用物资临时及适当地为伤病者进行的初步救援及护理，然后从速送院。

# 参考文献

[1] 王虹.美国通用航空的发展现状[J].中国民用航空, 2003, 000 (008): 45-47.

[2] 李寿平, 欧阳彦美.美国通用航空产业发展的法治经验及对中国启示[J].时代法学, 2015(1).

[3] 赵嶷飞, 涂堃.美国通用航空机场程序分析[J].中国民用航空, 2011 (10): 49-52.

[4] 耿建华.通用航空概论[M].北京: 航空工业出版社, 2007.

[5] 高启明, 金乾生.我国通用航空产业发展特征、关键问题及模式选择[J].经济纵横, 2013 (04): 104-108.

[6] 高启明."十三五"时期我国通用航空产业转型面临的挑战与发展思路[J].经济纵横, 2016 (2): 29-34.

[7] 李永明, 王俭勤, 郑晋光, 等.About Standard Avionics ATE of Oversease.国外标准化通用航空电子自动测试设备现状和发展[J].计算机测量与控制, 2004, 012 (001): 1-5.

[8] 董念清.A Research on the Situation, Difficulties and Countermeasures of China's General Aviation Development.中国通用航空发展现状、困境及对策探析[J].北京理工大学学报（社会科学版）, 2014, 016 (001): 110-117.

[9] 杨勇, 隋东.我国低空空域改革和通用航空事业发展有关问题的思考[J].南京航空航天大学学报: 社会科学版（2）: 50-53, 57.